Workbook to Accompany Modern Diesel Technology: Diesel Engines

Workbook to Accompany Modern Diesel Technology: Diesel Engines

SECOND EDITION

Sean Bennett

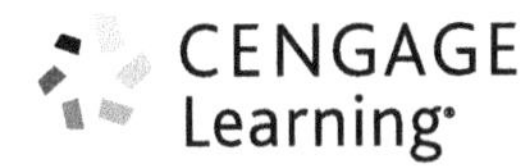

Australia • Brazil • Mexico • Singapore • United Kingdom • United States

Workbook to Accompany Modern Diesel Technology: Diesel Engines, Second Edition
Sean Bennett

VP, General Manager, Skills and Planning: Dawn Gerrain

Product Team Manager: Erin Brennan

Director, Development, Global Product Management, Skills: Marah Bellegarde

Senior Product Development Manager: Larry Main

Senior Content Developer: Sharon Chambliss

Product Assistant: Scott Royael

Marketing Manager: Linda Kuper

Market Development Manager: Jonathan Sheehan

Senior Production Director: Wendy Troeger

Production Manager: Mark Bernard

Content Project Manager: Christopher Chien

Art Director: Jackie Bates/GEX

Cover Image: Courtesy of Navistar, Inc.

Cover Inset Image: © 2015 Cengage Learning®; Photo Courtesy of Sean Bennett

Library of Congress Control Number: 2013950242

ISBN-13: 978-1-285-44298-3

ISBN-10: 1-285-44298-9

Cengage Learning
200 First Stamford Place, 4th Floor
Stamford, CT 06902
USA

Cengage Learning is a leading provider of customized learning solutions with office locations around the globe, including Singapore, the United Kingdom, Australia, Mexico, Brazil, and Japan. Locate your local office at: **www.cengage.com/global**

Cengage Learning products are represented in Canada by Nelson Education, Ltd.

For your course and learning solutions, visit **www.cengage.com**

Purchase any of our products at your local college store or at our preferred online store at **www.cengagebrain.com**

Printed at CLDPC, USA, 03-20

Table of Contents

Preface

STUDENT'S INTRODUCTION TO THE TEXTBOOK

Modern Diesel Technology: Diesel Engines was written as a classroom learning guide for entry-level students targeting a career as a diesel technician. The modern technician is required to have a theoretical understanding of diesel technology before attempting to work on engines on which the consequences of a mistake could be costly. This workbook addresses some basic shop-floor skills that relate to the theory in the core textbook. Anyone planning a career as a technician must have hands-on competence, but today, repair technology is so complex and changes so rapidly that all technicians must accept the need to rely on online service support systems. You may still find a few old-timers who will tell you that everything you need to know can be learned by hands-on practice, but you would do best to remind yourself that today's technician is not that of your father's generation.

FOCUS OF THE WORKBOOK

The textbook addresses basics. It introduces the fundamentals of diesel repair technology without detailing repair methods or techniques. The job sheets in the workbook are designed to help students tackle some of the routine tasks expected of rookie technicians in a diesel-repair facility. They are designed to be performed in a learning environment such as a high school or college. In some cases, the tasks may require equipment or data hub access not readily available in a learning environment. Students may still derive some benefit by reviewing these tasks and perhaps watching others perform them in a repair facility.

KEEPING CURRENT

Because technology advances so quickly, technicians tend to specialize more today than they did a generation ago. It is obviously easier to keep abreast of the changes that occur in one area of a technology than to try to maintain expertise on every chassis system. To repair today's equipment, training and frequent upgrading are required. Most companies today require ASE certification, if for no other reason than to minimize liability in the event of mechanical failures. Technicians today should make certification part of the goal of establishing a career as a vehicle repair expert.

OEMs are aware that regular training and updating of their technicians is required. It is not uncommon for technicians to attend several training courses per year. This type of training is usually geared to certified and experienced technicians. One objective of this textbook and its accompanying workbook is to help take the technician to the level of understanding required by the OEM to properly benefit from training courses designed for technicians.

ACHIEVING A PASSING GRADE

In school, college, and certification examinations we grade achievement by awarding either a letter or a percentage grade. In school we are taught that achieving an A or a grade of 80 percent or higher is the required standard. However, if you are planning to work on vehicles in any capacity, you are going to have to change what you think of as a passing grade. Performing a brake job that merits a score of 85 percent is not going to make your customer happy. In repair technology, every customer expects 100 percent. If you fail to achieve 100 percent, you will see that job back again, only this time you will be repairing it at no cost to the customer . . . and plenty of cost to your employer. This type of second-time-around repair is known as a "come-back" job. Come-backs cost everyone. They cost the customer because of increased downtime. They cost the employer of the technician. The bottom line is that an unhappy customer is not going to be paying for the time required to get the job done properly the second time around. In addition, the customer may not return with service work in the future.

Entry-level technicians will find it difficult to complete most shop-floor tasks in the OEM-specified times. As you are learning to master each technical procedure, first try to achieve the 100-percent grade required of each work task. Next, you can aim to meet the OEM book times.

CURRENCY

Diesel technicians may work in a variety of different fields. These include the following:

- Truck maintenance and repair
- Transit bus repair
- School bus repair
- Heavy equipment repair
- Stationary engine repair
- Marine engine repair
- Light-duty vehicle repair

Diesel engines have changed significantly during the last 10 years. This change has been driven by aggressive EPA and CARB legislation requiring engines that minimize emissions and produce better fuel economy (to lower greenhouse gases). All of the types of technology identified above change over time, but in no area is this change more pronounced than in highway trucks and buses. Many of the larger fleets maintain trucks through a life cycle that can be as short as three years. By doing this, their intent is to avoid most repairs and downtime. Often large fleets negotiate extended warranty plans with the OEMs. This allows the fleets to employ maintenance rather than repair technicians. A maintenance technician performs services and quick running repairs. Servicing a diesel engine consists of checking, topping up, or changing fluids such as the engine oil.

USING THE GLOSSARY

A comprehensive glossary is included at the end of the textbook. There is also a listing of acronyms. Make a practice of using both. Do not skip words you do not understand. You will soon find that your technical vocabulary develops and the result will be that you rely less on the glossary. As we advance into the electronic age, we tend to make increasing use of acronyms both in written and verbal communications. It is important that you become comfortable understanding acronyms.

The glossary is provided to help students interpret the technical vocabulary used in the textbook. For students who use English as a second language, the use of a good English dictionary, such as Webster's Collegiate, is recommended over one that interprets English to the foreign language. Students whose mother tongues are Spanish and French will find that many words in technology are anglicized in those and most other languages. Whether you like it or not, the reality is that the language of technology is English.

USING THE WORKSHEETS

The job sheets are designed so that they reflect some of the tasks that you might perform on the shop floor. As every school and college is differently equipped, the methods you use to complete the job sheets will vary. Some of the tasks require late-model, electronically managed engines. This makes access to OEM service information systems (SIS) and troubleshooting software desirable. However, if you have any OEM dealerships close by, they will often make the equipment available for demonstration or actually demonstrate it themselves. For the dealership it is simply good PR; they have an eye for potential and are always looking for future technicians. For the students it is exposure to technology they will eventually work with after they graduate.

The tasks have been made as generic as possible. Use the job training sheets as a guide and do not be afraid to adapt them for the technology you have at hand. However, some of the job sheets are necessarily OEM-specific. When you tune up a DDC-MUI-fueled engine, there are some distinct procedures that differentiate it from any other OEM diesel, so a DDC engine is obviously required. In each job sheet, a list of tools is provided. This does not include hand tools, which are a requirement for any service procedure.

USING FUNCTIONAL ENGINES

There is probably nothing as damaging to the development of sound technical practice in students than working on engines that are never expected to run again. The possible exception is the engine that is completely disassembled and used solely for parts inspection, measuring, and failure analysis. When a technician is being trained in tune-up procedure, for example, if part of the procedure is to performance-test the engine after the tune-up, the student will automatically work to a higher standard. Shortcomings in attaining the required standard will be identified immediately, and the remedies required to correct the shortcoming will further assist in the development of sound performance standards.

Stripping down engines with 25 percent of the parts missing or defective results in a negative teaching and learning experience for students. If you recognize that the work you are performing is futile, it is human nature to care little about the end result. If only "dead" engines are available in a training facility, avoid strip-down and reassembly procedures and replace them with short, set exercises with a clearly defined objective such as *Recondition a set of cylinder valves.*

WORKSHOP PRACTICE

Sound job organization and housekeeping are just as important in a college training environment as in industry. A messy, disorganized shop says something to anyone visiting it—either as a customer or as a worker. The impression is a negative one, even in the unlikely event that the work performed in such an environment is satisfactory.

As a novice technician, you should make it your business to develop good workshop practice habits. Observe the working methods of competent, experienced technicians. You can often learn more just by watching and listening than by asking dozens of questions. When you disassemble components, the methods you use should be guided by the service literature. Your workbench should be clean and uncluttered, with disassembled parts organized by system in trays and cans. Make a habit of labeling components. When you use shop equipment, leave the workstation completely clean after completion of a job.

IMPORTANCE OF OEM SERVICE LITERATURE AND SOFTWARE

In this age it is necessary to emphasize the importance of using the OEM procedure, which means using the OEM service literature or software when following any shop procedure, however simple. Make a habit of reading technical service literature. More importantly, make sure you fully understand what you are reading—remember that advice about using the glossary. It becomes very difficult to precisely execute a sequential troubleshooting table if the text at each step is not properly understood. Work at developing your reading skills. This will not only help you to navigate OEM service literature but it can also help you with the self-directed programs of learning that the industry is increasingly using. When you complete work orders and job sheet profiles, make sure that the words you use are concise and accurately convey the message you are trying to communicate.

ELECTRONIC MEDIA

Many OEMs make their diagnostic software and data hub access available to training institutions because it is in their interest to have knowledgable technicians working on their products. Students today must get used to working with electronic information, and it is important to be exposed to networking technology in a manner that is not intimidating and has some relevance to shop-floor tasks. Once the basics of Internet protocols are understood, it is recommended that students be assigned specific tasks to accomplish while networking. Even if Mitchells', All-Data, SAE software, or proprietary data hub network access are not available, the Internet can be used to hunt down information on most shop procedures, component parts (especially oddball parts that are not easily available through more conventional channels), and technical data on current equipment.

SUPPORT RESOURCES

The typical high school or college resource center tends to be academic and of limited use to truck technicians. Truck technician training facilities should be equipped with networked computers, a good selection of OEM service literature, a video library, and a good selection of textbooks that cover the theoretical requirements of the program of study. Students should also refer to some texts that challenge them at a higher level than the required standards of the course outline. See if you can obtain subscriptions to a selection of trade magazines. Get used to using a variety of sources for information rather than relying on a single source, especially when undertaking research projects.

SUGGESTIONS FOR REFERENCE MATERIALS

Recommended Practices Manual Technology and Maintenance Council of the ATA

Machinery's Handbook (Industrial Press)

Webster's Collegiate Dictionary (Merriam Webster)

Heavy Duty Truck Systems, fifth edition, Bennett (Cengage)

Electricity and Electronics, second edition, Bell (Cengage)

Electronic Diesel Engine Diagnosis, Bennett (Cengage)

Glossary of Automotive Terms (SAE)

Dictionary of Automotive Technology, Bennett (Centennial College Press)

Automotive Dictionary, South and Dwiggens (Delmar ITC)

Truck Engines, Fuel, and Computerized Management Systems, fourth edition, Bennett (Cengage)

BEST TRADE MAGAZINES

Free online subscriptions are available for many trade magazines today, including the first two in this list.

CCJ (Commercial Carrier Journal) (etrucker.com)

Diesel Progress (dieselprogress.com)

SAE Automotive Engineering

Motor Magazine

Heavy Duty Trucking

Transportation Topics (ATA monthly magazine)

TEST PREPARATION

Get used to the idea of writing and passing written tests! Whether you like it or not, to pass college courses or ASE certification you will have to take written tests. Do not wimp out and use the excuse that you have no problem with hands-on competency but you are hopeless at tests. Today's technician is literate and must have the ability to pass multiple-choice tests.

Every student learns differently, so the learning method that works best for one person may not work for another. Try to determine which study methods work best for you. Generally, you should not leave review and study time to the last moment. Spending 15 minutes a day for six months will be more effective with most learners than spending a week before a test studying to the point of exhaustion. Most learners forget material if they do not review the subject matter. Look at the following data produced from testing following a lecture:

% Retained in Memory	Without Review	With Review
After 24 hours	60	80
After one week	50	75
After two weeks	40	70
After three weeks	30	70
After four weeks	20	70

The preceding chart shows the value of establishing a routine of review. For most learners, reviewing material is a key to succeeding in exams. The great thing about making a review a habit is that it does not take long. If you properly understood the material the first time around, minutes are all that is required to refresh the learning experience in your mind. The following is an example of a review timetable that will work for most students:

REVIEW 1

This should take place within 24 hours of the learning experience. The same day works best for most learners. This review step is the most important in the review process. Use point-form notes; that is, keep notes as brief as possible. When instructors write out notes or issue handouts, abbreviate them. Short-form notes are a great study tool; just make sure that they are not abbreviated to the extent that they cease to have meaning to you. In this first review, try to understand the subject matter. Make notes about material you do not grasp and make it your business to get those questions answered either by private research or by your teacher.

REVIEW 2

Go over your abbreviated notes in this session. Make sure you understand both your own notes and the subject matter covered. This second review should take a matter of minutes; you will be recalling the learning experience and the time you devoted to the first review.

REVIEW 3

In this step, you simply relive the initial learning experience and the first review step. It should take little time, and you may even feel it is boring. However, it is a great way to reinforce the information learned. Depending on what type of learner you are, you may want to do this once a month. But do it—it does work, and the payoff will come when you are tested.

REVIEW 4

This should be done immediately before testing. Most learners do not retain information well when their only study occurs immediately prior to an exam. Try not to spend too long studying because if you do, your brain will be exhausted by the time you have to take the test. One thing that can work well is joining a small study group before a test. Use sets of typical test questions, such as those found in this book, as a basis for discussion.

ACTIVE LEARNING

Mechanical technicians tend to learn best by doing rather than thinking, which is why we often feel out of place in classrooms. Become active in the classroom by using some of the following strategies:

1. *Make your own notes.* Most teachers write far too much on boards and in their handouts, so rewrite what they say in terms that mean something to you. Notes are for you only. Challenge yourself to make them as meaningful as possible. Remember, notes are a great review tool, especially if you can capture the contents of a three-hour class in one page of bullet-form notes.
2. *Draw diagrams.* Try not to always rely on handouts. Drawing a diagram makes you active in the class. Even if you do not have much artistic ability, the actions required to draw a diagram will usually help you better understand and remember the technology.
3. *Ask questions in class.* That is, ask them if you feel comfortable doing so. Not all learners like to ask questions in a formal class. But asking questions, even if you think are they dumb questions, is a way of making yourself an active learner, which helps retention.
4. *Look for ways of making connections between the theoretical information you get in class and the hands-on practice of repair technology.* This works well if you have worked hands-on with the technology you are learning in class.

READING THE TEXTBOOKS

Most textbooks are difficult to pick up and read in serial fashion like a novel. When you study directly from the textbook, define your goals before opening the book, then use the book to meet them. Before you begin, it is a good idea to know something about the goals you want to achieve, so write them down in brief form. Next, you have to navigate the contents of the textbook to achieve those goals. Here is an action plan I use when studying subject matter that is new to me:

1. *Goals.* Define your goals. Write them down on paper. They could be as simple as a couple of words or more complex, depending on what it is you are expecting to learn.
2. *Survey.* You have your goals in note form on paper. Next, consult the textbook table of contents to determine if the book is going to be of any help. Try to avoid using the index to search for information; indexing does not necessarily indicate where critical explanations occur in the text. Once you have targeted a chapter in the textbook, take a look at the chapter objectives. This should give you a pretty good idea of whether you are on track or not.
3. *Read, distill, and restate.* Target the information you need in the text. Read the content and get a general sense of what the author is trying to say. In most cases what the author is saying can be restated in many fewer words. Using pen and paper, take bullet notes on the content. Keep it brief. Most importantly, put everything in words that make sense to you.
4. *Review.* Close the textbook and go over the bullet-form notes you have made. Does the information make sense to you? If it does not, you will have to open the book up again and redo the notes. If it does make sense and the notes you have made are sufficient to allow you to recall the material later on, you have achieved your objective.

IMPROVING RETENTION

Because we all learn differently, remember that what works for you may be different from what works for others. We said before that mechanical technicians generally learn best when a learning experience is active. That often means hands-on learning, which is not always possible in college programs with congested curricula and large classes, so look for other ways of making the learning experience real for you. Here are some key things that might work:

- *Active classroom learning.* Use the classroom techniques we described earlier to avoid being a completely passive participant in the learning process. Because you are more likely to learn by doing, DO as much as possible in class. Condense existing notes, draw diagrams, ask questions—even the dumb ones.
- *Research information.* Use the Internet if this approach to learning works for you. It is a great way of supplementing what you have learned in the classroom, and the best thing is that you control both the learning path and the pace. Again, make short bullet-form notes when you are hunting down information online.
- *Use video.* Most colleges have video libraries that are too lengthy to run in structured classes. But if you like to learn visually, use the video libraries in resource centers to learn. Most OEMs make good-quality digital video, some of it interactive. Note: Video is a big turn-off for some and has a way of inducing sleep, especially after lunch. If this is true for you, recognize it and target short, to-the-point video presentations.
- *Active research.* This can really be effective. In teaching engine technology, it is common to hear students complain that they never work on engines and find it difficult to understand repair techniques. Most technicians should be able to obtain a discarded engine for little or no investment, whether it is a diesel or not. Take one apart in your garage and reassemble it. It doesn't matter whether it will ever run again; the experience of disassembly and reassembly will stick in your mind forever. Just do not make a habit of working on a "dead" engine. After this initial experience, make sure you graduate to working on engines that are expected to run again.
- *Be curious.* This is your career. Do not black-box technology. If you do not understand how a particular component functions, get an example that has failed, test it with test instruments, and take it apart with a hacksaw if necessary. Make it your business to answer those questions that no one seems to have an answer for.

IMPROVING TEST-TAKING SKILLS

Most truck technicians are not academics by inclination, and it often seems unfair that they are examined academically rather than by practical tasks. The modern automotive technician must be literate, so attempting to justify a paper examination failure by claiming to have mastered all the hands-on competencies fools no one but yourself. However, there are some simple things you should be aware of that can greatly improve your ability to succeed on tests:

1. The number one reason for failing a test is simply not knowing the material. Use the study techniques we describe in this workbook and the textbook to ensure that you understand the subject matter.
2. Never spend too much time analyzing a test question. If you do not understand a question, skip it and return to it when you have completed the questions you do understand. Analyzing is a great skill for a technician—but it can hinder you when writing a test by causing you to read meanings into a test question that are not there.
3. A typical college or ASE test question consists of a stem or question followed by four possible answers. Only one of the four possible answers is correct; the other three are known as *distractors*.
4. Read with your pencil. Underline key words in the question. Eliminate distractors that make no sense at all. If you have to guess, make sure your guess is an intelligent one; you can only do this if you have already eliminated the distractors that do not make sense.
5. Read the question and ALL the answers. In a multiple-choice test, you are selecting the most correct answer; it may be that some of the distractors are in some way correct.
6. Distribute your time appropriately. This is especially important if you know you write tests slowly. Take off your watch and place it above the test paper. Divide the test into sections. Answer those questions you find easy, first.
7. Forget about answer patterns. It is of no significance if you have answered question A for the previous four questions. Most tests are drawn from computerized test question banks, so there is nothing significant about the correct answer patterns expected in the test.
8. Think twice before changing an answer in a multiple-choice test. Studies show that more often than not a correct answer is changed to an incorrect answer—take a look at item 2 about overanalyzing test questions. The answer that first occurs to you is likely to be correct.
9. Erasing. Most tests today are graded optically by scanning. If you erase an answer, make sure that it is completely erased. If not, two answers will be scanned for the question and you will get the question wrong. If you are unable to properly erase a response that you feel is incorrect, ask for another answer sheet.
10. Answer every question. In most tests there are four possible answers for every question and only one answer is correct. You do not get penalized for incorrect responses. Even if you are entirely guessing the answer, you still have a 25 percent chance of getting it right.
11. Relax! If the test becomes confusing to you, spend five minutes daydreaming to get your mind off the test. You may find that what was confusing becomes less so. Breathe slowly and deeply if you are inclined to panic during tests.
12. This is a tough one to swallow: Do not expect every question in the test to be absolutely technically accurate. Test questions are written by human beings, usually by technicians who as a rule are not academics. It is possible for the odd question to appear in a test that makes little technical sense. If you see a question like this, try asking yourself which answer the author of the question might have thought was correct. This might be easy if the author is your instructor in a college program, or a lot harder if you are taking a certification test. More importantly, avoid getting too worked up over one bad question! Move on to the next.

Be positive. Tell yourself that you are going to pass any test you take no matter what. Just make sure you give yourself a fair chance by properly preparing for it ahead of time.

CHAPTER 1

Shop and Personal Safety

OBJECTIVES

After studying this chapter, you should be able to:

- **Identify potential danger in the workplace.**
- **Describe the importance of maintaining a healthy personal lifestyle.**
- **Outline the personal safety clothing and equipment required when working in a service garage.**
- **Distinguish between different types of fire.**
- **Identify the fire extinguishers required to suppress small-scale fires.**
- **Describe how to use jacks and hoisting equipment safely.**
- **Explain the importance of using exhaust-extraction piping.**
- **Identify what is required to work safely with chassis electrical systems and shop mains electrical systems.**
- **Outline the safety procedures required to work with oxyacetylene torches.**

END OF CHAPTER REVIEW QUESTIONS

1. When is a worker most likely to be injured?
   a. First day on the job
   b. During the first year of employment
   c. During the second to fourth year of employment
   d. During the year before retirement
2. When lifting a heavy object, which of the following should be true?
   a. Keep your back straight while lifting
   b. Keep the weight you are lifting close to your body
   c. Bend your legs and lift using the leg muscles
   d. All of the above
3. Technician A says that a B-class fire should be extinguished with an A-category fire extinguisher. Technician B says than an electrical fire should be suppressed with a C-category fire extinguisher. Who is right?
   a. Technician A only
   b. Technician B only
   c. Both A and B
   d. Neither A nor B

4. Which of the following is usually a requirement for a safety shoe or boot?
   a. UL certification
   b. Steel sole shank
   c. Steel toe
   d. All of the above
5. What type of glove should NEVER be worn when working with refrigerants?
   a. Synthetic rubber gloves
   b. Vinyl disposable gloves
   c. Leather welding gloves
   d. Latex rubber gloves
6. Which of the following is under the most pressure?
   a. Oxygen cylinders
   b. Acetylene cylinders
   c. Diesel fuel tanks
   d. Gasoline fuel tanks
7. Which type of fire can usually be safely extinguished with water?
   a. Class A
   b. Class B
   c. Class C
   d. Class D
8. When attempting to suppress a Class C fire in a chassis, which of the following is good practice?
   a. Disconnect battery power
   b. Use a carbon dioxide fire extinguisher
   c. Avoid inhaling the fumes produced by burning conduit
   d. All of the above
9. What color is used to indicate the fuel hose in an oxyacetylene station?
   a. Green
   b. Red
   c. Yellow
   d. Blue
10. In which direction do you tighten an oxygen cylinder fitting?
    a. Clockwise (CW)
    b. Counterclockwise (CCW)
    c. Depends on the manufacturer

JOB SHEET 1.1

Name ______________________ Date ______________

Job Description: Introduction to shop safety equipment.

Performance Objective: After completing this assignment, the student should be able to use shop tools and lifting equipment in a safe manner, and be aware of the consequences of incorrect or unsafe work methods.

Text Reference: Chapter 1.

Protective Clothing: Required for shop exercise sessions: hearing, eye, hand, and foot protection, and shop coat/coveralls.

Tools and Materials: Any shop equipped to remove diesel engines from a chassis and rebuild them. This could be a college shop, general service garage facility, or specialty engine overhaul shop.

PROCEDURE

1. Student to produce a floor map of the shop and identify the locations of safety exits. List the locations.
   - ______________
   - ______________
   - ______________

   Task completed ______________

2. Student to identify the correct steps and procedures involved in shop or classroom evacuation.
   - ______________
   - ______________
   - ______________

   Task completed ______________

3. Student to locate recommended fire extinguishers and identify them by type. List the locations.
   - ______________
   - ______________
   - ______________

   Task completed ______________

4. Student to locate vehicle- and component-lifting and wheel-blocking devices within the shop area. List the number of available units and note when last tested.
   - Jack stands ______________
   - Engine hoists ______________
   - Wheel chocks ______________
   - Overhead crane(s) ______________

- Lift truck(s) ________________
- Floor jack(s) ________________
- Oxyacetylene stations ________________
- Other ________________

Task completed ________________

5. Student to locate "shop tool area" and note condition of:
   - Bench grinder safety guards ________________
   - Oxyacetylene torches/tips, gauges/covers ________________
   - Gas tank storage area ________________
   - Oxygen gas storage area ________________
   - A/C gas storage area ________________
   - Propane gas (lift truck fuel) storage area ________________
   - High-pressure wash equipment, check hoses ________________
   - Safety shields (gas/arc/MIG and grinding) ________________
   - Vises (worm gear, jaws) ________________
   - Breathing apparatus/ventilation ________________

Task completed ________________

Instructor Check/Comments

____

____

____

JOB SHEET 1.2

Name ______________________________ Date ______________

Job Description: Familiarization with ANSI symbols.

Performance Objective: After completing this assignment, the student should be able to interpret common ANSI symbols on shop equipment and supplies.

Text Reference: Chapter 1.

Protective Clothing: Required for shop exercise sessions: hearing, eye, hand, and foot protection, and shop coat/coveralls.

Tools and Materials: Any shop equipped to undertake repairs on mobile equipment and diesel engines. This could be a college shop, a general service garage facility, or specialty engine overhaul shop.

PROCEDURE

Refer the ANSI safety symbols in **Figure 1–1** and identify by name any equipment and products on the shop floor on which these symbols appear.

This symbol indicates *prohibited action*

Do not overtorque: do not use pipes or lever extensions

No hammering

No prying

Do not strike hard objects

No impact/power drive

Do not step in or on drawers

Do not open multiple drawers

Do not pull to move

This symbol indicates *mandatory action*

Mandatory ear protection

Mandatory face shield

Mandatory mask

Mandatory respirator

Mandatory protective clothing

Mandatory protective gloves

Mandatory eye protection

Must read instructions before use

This symbol indicates a *hazard alert*

Red background—danger
Orange background—warning
Yellow background—caution

Vibration hazard

Risk of explosion

Overhead/overload hazard

Risk of electric shock

Risk of fire

Risk of entanglement

Courtesy of Snap-on Tools Company

FIGURE 1–1 ANSI Safety Symbols.

Symbol	Describe Product or Equipment
1.	
2.	
3.	
4.	
5.	
6.	
7.	
8.	
9.	
10.	
11.	
12.	

Task completed ______________

Instructor Check/Comments

____

____

____

INTERNET TASKS

Search the Internet for information on the environmental and safety concerns regarding personal and shop equipment. Type the following items into your search engine to get started:

1. Red Wing® safety footwear
2. SSI® safety glasses
3. 3M™ respirators
4. CSM® industrial safety products
5. Maximus™ electronic earmuffs
6. SnugPlugs™ (hearing protection)
7. ULINE® fire extinguishers
8. Safety Supply Company®

STUDY TIPS

Identify five key points in Chapter 1.

Key point 1 ______________________________

Key point 2 ______________________________

Key point 3 ______________________________

Key point 4 ______________________________

Key point 5 ______________________________

CHAPTER 2

Hand and Shop Tools, Precision Tools, and Units of Measurement

OBJECTIVES

After studying this chapter, you should be able to:

- Identify the hand tools commonly used by diesel technicians and describe their functions.
- Categorize the various types of wrenches used in shop practice.
- Describe the precision measuring tools used by the engine and fuel system technician.
- Outline the operating principles of a standard micrometer and name the components.
- Identify different types of torque wrenches.
- Calculate torque specification compensation when a linear extension is used.
- Read a standard micrometer.
- Outline the operating principles of a metric micrometer and name the components.
- Read a metric micrometer.
- Understand how a dial indicator is read.
- Define TIR and understand how it is determined.
- Understand how a dial bore gauge operates.
- Outline the procedure for setting up a dial bore gauge.
- Perform accurate measurements using a dial bore gauge.
- Describe some typical shop hoisting equipment and its applications.

END OF CHAPTER REVIEW QUESTIONS

1. When the spindle contacts the anvil on a standard 0- to 1-inch micrometer, it should read
   a. zero.
   b. one thousandth of an inch.
   c. 0.025 inch.
   d. 1 inch.

2. How many complete rotations must the thimble of a standard micrometer be turned to travel from a reading of zero to a reading of 1 inch?
   a. 25
   b. 40
   c. 50
   d. 100
3. How many complete rotations must the thimble of a standard metric micrometer be turned to travel from a reading of zero to a reading of 25 millimeters?
   a. 25
   b. 40
   c. 50
   d. 100
4. When the thimble of a metric micrometer is turned through one full revolution, the dimension between the anvil and the spindle has changed by
   a. 0.1 millimeters.
   b. 0.5 millimeters.
   c. 2.5 millimeters.
   d. 5.0 millimeters.
5. When using a dial indicator to check the concentricity of a flywheel housing, during a single rotation of the flywheel, the reading on the positive side of the zero on the dial peaks at 0.002 inch, while the reading on the negative side peaks at 0.005 inch. What is the TIR?
   a. 0.003 inch
   b. 0.005 inch
   c. 0.07 inch
   d. 0.010 inch
6. Which of the following precision measuring instruments would be required to measure a valve guide bore?
   a. Dial indicator
   b. Inside micrometer
   c. Small hole gauge and micrometer
   d. Dial bore gauge
7. How many radial strokes on the cap screw head indicate that a bolt is an SAE grade 5 bolt?
   a. 3
   b. 5
   c. 6
   d. 8
8. Technician A says that impact sockets are usually manufactured from harder steel than regular sockets. Technician B says that an impact socket is designed for use exclusively on impact guns. Who is correct?
   a. Technician A only
   b. Technician B only
   c. Both A and B
   d. Neither A nor B
9. When drilling cast iron, the correct method calls for the procedure to be performed
   a. with kerosene.
   b. with lard.
   c. preheated.
   d. dry.
10. Which Plastigage color code should be selected to measure a main bearing clearance that the manufacturer specifies must be between 0.0023 inch and 0.0038 inch?
    a. Red
    b. Green
    c. Blue

INTERNET TASKS

Use a search engine to access the following tool manufacturers and suppliers. Check out any online deals!

1. Snap-on® tools
2. Mac Tools®
3. Craftsman® tools
4. Starrett® tools
5. Mitutoyo® tools
6. Blackhawk™ tools
7. Proto® tools
8. Armstrong® tools
9. Mastercraft™ tools
10. Gray Tools®

STUDY TIPS

Identify five key points in Chapter 2.

Key point 1 ______________________________

Key point 2 ______________________________

Key point 3 ______________________________

Key point 4 ______________________________

Key point 5 ______________________________

CHAPTER 3

Engine Basics

OBJECTIVES

After studying this chapter, you should be able to:

- **Interpret basic engine terminology.**
- **Identify the subsystems that make up a diesel engine.**
- **Calculate engine displacement.**
- **Define the term *mean effective pressure*.**
- **Describe the differences between a *naturally aspirated* and a *manifold-boosted* engine.**
- **Explain how volumetric efficiency affects cylinder breathing.**
- **Define *rejected heat* and explain thermal efficiency in diesel engines.**
- **Outline the operation of a diesel four-stroke cycle.**
- **Outline the operation of a diesel two-stroke cycle.**
- **Calculate engine displacement.**
- **Interpret the term *cetane number* and relate it to ignition temperature.**

END OF CHAPTER REVIEW QUESTIONS

1. A diesel engine has a bore of 4.5 inches and a stroke of 5.25 inches. Which of the following correctly describes the engine?
   a. 10-inch displacement
   b. 10-liter displacement
   c. Oversquare
   d. Undersquare
2. Which of the following best describes the term *engine displacement*?
   a. Total piston swept volume
   b. Mean effective pressure
   c. Peak horsepower
   d. Peak torque
3. Engine breathing in the two-stroke cycle is sometimes referred to as
   a. inertial.
   b. scavenging.
   c. exhaust blowdown.
4. The tendency of an object in motion to stay in motion is known as
   a. kinetic energy.
   b. dynamic friction.
   c. inertia.
   d. mechanical force.

5. What percentage of the potential heat energy of the fuel does the modern diesel engine convert to useful mechanical energy when it is operating at its best efficiency?
   a. 20 percent
   b. 40 percent
   c. 60 percent
   d. 80 percent
6. Where does scavenging take place on a four-stroke cycle diesel engine?
   a. BDC after the power stroke
   b. TDC after the compression stroke
   c. Valve overlap
   d. 10 to 20 degrees ATDC on the power stroke
7. Ideally, where should peak cylinder pressure occur during the power stroke if it is to deliver torque to the flywheel as smoothly as possible?
   a. TDC
   b. 10 to 20 degrees ATDC
   c. 90 degrees ATDC
   d. At gas blowdown
8. When running an engine at any speed or load, which of the following would be the best location to produce peak cylinder pressure?
   a. 20 degrees BTDC
   b. 10 degrees BTDC
   c. TDC
   d. 15 degrees ATDC
9. As a piston descends through its power stroke, at which point would the crank throw leverage be at its highest?
   a. TDC
   b. 15 degrees ATDC
   c. 90 degrees ATDC
   d. BTDC
10. The energy of motion is known as
    a. inertia.
    b. kinetic.
    c. thermal.
    d. potential.

JOB SHEET 3.1

Name ______________________________ Date ______________

Job Description: General orientation of a commercial diesel engine.

Performance Objective: After completing this assignment, the student should be able to understand the basics of diesel engine operation, connect key terms to engine hardware, and identify the events of the four- and two-stroke diesel cycles.

Text Reference: Chapter 3.

Protective Clothing: Required for shop exercise sessions: hearing, eye, hand, foot protection, and shop coat/coveralls.

Tools and Materials:

Standard shop tools
A diesel engine cutaway (preferable) that can be rotated, or an assembled diesel engine
Pencil, paper, textbook, and OEM service literature if available

PROCEDURE

1. Identify the location of intake and exhaust system components. Identify the impeller and turbine housings of the turbocharger and follow gas flow routing from the air cleaner to the exhaust after treatment canister.
   - ______________
   - ______________
   - ______________

Task completed ______________

2. Identify the serviceable components on the cooling and lubrication circuits. This should include filters, heat exchangers, oil fill and sump drain locations, etc.
   - ______________
   - ______________
   - ______________

Task completed ______________

3. Identify the key engine data.
   - Bore ______________
   - Stroke ______________
   - Swept volume ______________
   - Clearance volume ______________
   - Compression ratio ______________
   - Peak torque ______________
   - Rated power ______________
   - Engine firing order ______________

Task completed ______________

4. Is the engine:
   - Undersquare ________________
   - Oversquare ________________
   - Square ________________

Task completed ________________

5. Identify the diesel fuel used and research some fuel data (this information may be difficult to locate; contact fuel supplier).
   - Type of fuel (ULS 1D or 2D) ________________
   - Maximum sulfur content ________________
   - CN rating ________________
   - Ignition temperature ________________

Task completed ________________

6. Identify the seven subcircuits of the engine you are working on and note anything distinctive.
   - Engine housing ________________
   - Engine powertrain ________________
   - Engine camshaft and valve train ________________
   - Engine lube circuit ________________
   - Engine cooling circuit ________________
   - Engine breathing circuit ________________
   - Engine (fuel) management system ________________

Task completed ________________

Instructor Check/Comments

____

____

____

JOB SHEET 3.2

Name ______________________________ Date ______________

Job Description: Identify a diesel engine by accessing the chassis data bus.

Performance Objective: Connect an electronic service tool (EST) to the chassis data bus with the objective of recording specification data and service history. NOTE: This task is usually the first step taken when a truck is brought into a shop—for students to understand what they are doing, they may need to refer to Chapter 14.

Text Reference: Chapters 3 and 14.

Protective Clothing: Required for shop exercise sessions: hearing, eye, hand, and foot protection, and shop coat/coveralls.

Tools and Materials:

A truck, transit bus, or off-highway heavy equipment with a chassis data bus
An EST and appropriate OEM software or HD reader capability
A communications adaptor and data bus connector
A means of recording the data

PROCEDURE

The procedure you use will depend on what type of equipment you can access and the level of access afforded by the software you are using. **Figure 3–1** shows how this would be done on a truck chassis. Note that the EST used in this example is a laptop computer and that a communications adaptor (CA) is required.

1. Boot the EST you are using. If it is PC-based, launch the OEM software.

   Task completed ______________

2. Connect the EST to the data bus connector.

   Task completed ______________

3. Access the chassis data bus.

   Task completed ______________

4. Record the engine specification data.

   Task completed ______________

5. Record the service/repair history log. Depending on the specific chassis electronics, you may have to exit the engine portal and enter a separate chassis service log.

   Task completed ______________

This exercise may be repeated a number of times on different vehicles, allowing you to build a file collection to be used for comparative purposes.

Instructor Check/Comments

____

____

____

PS3-1 Location of a 9-pin data connector on a Volvo Class 8 chassis: underside of the dash, left of the steering column.

PS3-2 Data bus can be accessed using a ProLink 9000 and the appropriate cartridge.

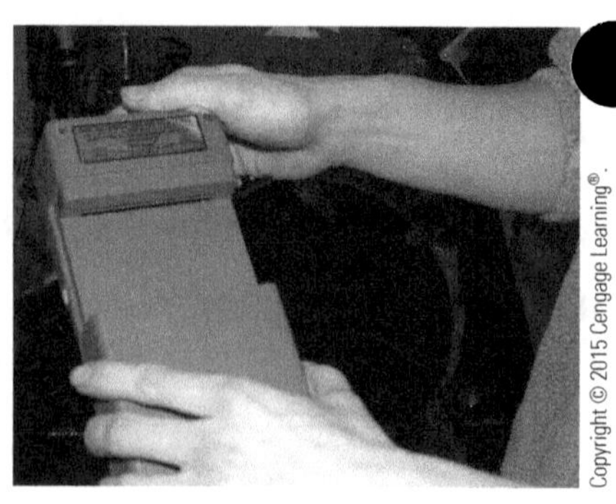

PS3-3 To install the cartridge into the ProLink head, snap it into position using a firm forward thrust.

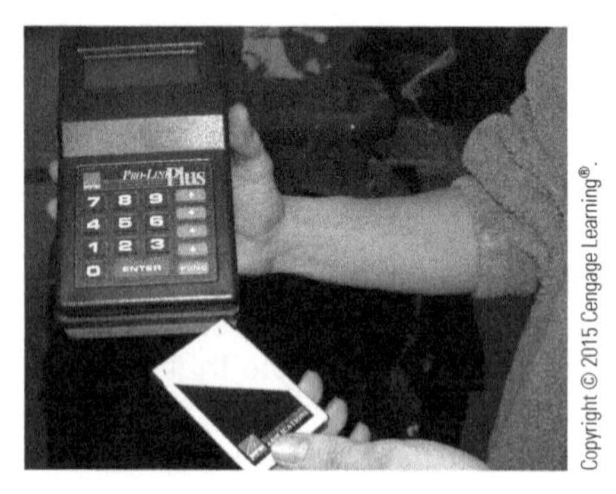

PS3-4 Alternatively, a ProLink MPC may be used: The technician is shown inserting a data card into an MPC cartridge.

PS3-5 Technician preparing to make a handshake connection from an electronic service tool (EST) to a J1939 chassis data bus: Note that this should be locked into position with a CW twist. Verify the connection by checking that the LED is illuminated on the CA.

PS3-6 Technician using a portable computer station loaded with diagnostic software after the handshake connection has been made. The PC screen shows the graphical display at the point of reading a chassis data bus.

FIGURE 3–1 Accessing a data bus with an HH-EST and PC.

INTERNET TASKS

Use a search engine to access the following websites and identify the current engine families of the following OEMs.

1. Caterpillar®
2. Cummins®
3. Detroit Diesel®, Mercedes-Benz®
4. Volvo-Mack®
5. Navistar® International Engine Corp.
6. Paccar®
7. Hino Trucks™, Toyota®

STUDY TIPS

Identify five key points in Chapter 3.

Key point 1 ______________________________

Key point 2 ______________________________

Key point 3 ______________________________

Key point 4 ______________________________

Key point 5 ______________________________

CHAPTER 4

Piston Assemblies, Crankshafts, Flywheels, and Dampers

OBJECTIVES

After studying this chapter, you should be able to:

- Identify the engine powertrain components.
- Define the roles of piston assemblies, crankshafts, flywheels, and dampers.
- Identify the different types of pistons used in current diesel engines.
- Describe the combustion chamber designs used in diesel engines.
- Explain the function of piston rings.
- Classify piston wrist pins by type.
- Describe the role of connecting rods, and outline the stresses they are subjected to.
- Identify common crankshaft throw arrangements.
- Outline the forces a crankshaft is subjected to under normal operation.
- Identify some typical crankshaft failures and their causes.
- Outline the procedure for an in-chassis rod and main bearing rollover.
- Measure friction bearing clearance using Plastigage.
- Define the term *hydrodynamic suspension.*
- Outline the roles played by vibration dampers and flywheel assemblies.
- Describe how vibration dampers function.
- Remove and replace a ring gear on a flywheel.

END OF CHAPTER REVIEW QUESTIONS

1. Where would you likely find a Ni-Resist insert?
   a. Under right belt, aluminum trunk piston
   b. Lower ring belt, articulating piston
   c. Upper ring belt, crosshead piston
   d. Trunk piston, pin boss
2. Under which of the following conditions would a piston ring seal most effectively?
   a. Low engine temperatures
   b. High lube oil pressures
   c. Low cylinder pressures
   d. High cylinder pressures
3. Which of the following crankshaft journal surface hardening methods provides the highest machinability margin?
   a. Nitriding
   b. Shot peening
   c. Flame hardening
   d. Induction hardening
4. What should you use to measure rod and main journal bearing clearances?
   a. Tram gauges
   b. Dial indicators
   c. Plastigage
   d. Snap gauges
5. What type of piston is favored by diesel engine OEMs in recent commercial vehicle diesel engines?
   a. Aluminum trunk
   b. Forged steel trunk
   c. Two-piece articulating
   d. Two-piece crosshead
6. Which of the following types of pistons uses bushingless pin bosses?
   a. Crosshead
   b. Ferrotherm
   c. Monotherm
   d. Aluminum alloy
7. Technician A says that keystone rings are the most common type used in current diesel engines. Technician B says that keystone conrods have a wedge-shaped small end. Who is correct?
   a. Technician A only
   b. Technician B only
   c. Both A and B
   d. Neither A nor B
8. Which of the following would be most likely to occur if an engine was run under load with a failed vibration damper?
   a. Increased fuel consumption
   b. Increased cylinder blowby
   c. Camshaft failure
   d. Crankshaft failure
9. What is the upper face of a piston assembly known as?
   a. Crown
   b. Skirt
   c. Boss
   d. Trunk

10. Which term is used to describe the cylinder volume between the piston upper compression ring and its leading edge?
    a. Dead volume
    b. Headland volume
    c. Toroidal recess
    d. Clearance volume
11. Technician A says that when using Plastigage to measure main bearing clearance, the larger the squish dimension, the greater the clearance. Technician B says that when measuring main bearing clearance it is important to rotate the engine through one full revolution after clamping the Plastigage strip into position. Who is correct?
    a. Technician A only
    b. Technician B only
    c. Both A and B
    d. Neither A nor B
12. When you remove a main bearing cap after checking bearing clearance with the specified color-coded Plastigage, you find that the nylon strip has failed to deform. Which of the following is likely to be true?
    a. Excessive bearing clearance
    b. Insufficient bearing clearance
    c. Warped crankshaft
    d. Excessive main cap clamping pressure

JOB SHEET 4.1

Name ______________________ Date ______________

Job Description: Crankshaft inspection.

Text Reference: Chapter 4.

Performance Objective: After completing this assignment, the student should be able to inspect a crankshaft to a set of OEM service specifications.

Protective Clothing: Required for shop exercise sessions: hearing, eye, hand, and foot protection, and shop coat/coveralls.

Tools and Materials:

Crankshaft from a six- or eight-cylinder diesel engine
V-blocks
Outside micrometers sized to measure rod and main journals
OEM service specifications

PROCEDURE

Visual Inspection

1. The visual inspection of a crankshaft is as important as measuring it. Mount in appropriately sized V-blocks to enable measurement to be made; the straightness of the crankshaft may be measured in V-blocks, but because a crankshaft is designed to bend, make sure that it is supported on the journals recommended by the OEM to perform this test.
2. Check for cracks and heat checking (bluing) at each journal.
3. Check for visible cracks at each rod and main journal fillet.
4. Check for visible cracks at each journal oil hole.
5. Check the thrust faces for scoring and wear.
6. Check each oil seal race for wear.

Tasks completed ______________

Measuring the Crankshaft

1. Once the crankshaft has passed a visual inspection it can be measured with the micrometers.
2. Out-of-round measurements: Measure each journal in two longitudinal locations and compare with a 90-degree offset measurement.

Main journal 1 1 ____________ 2 ____________

1 ____________ 2 ____________

Main journal 2 1 ____________ 2 ____________

1 ____________ 2 ____________

Main journal 3 1 ____________ 2 ____________

1 ____________ 2 ____________

Main journal 4 1 ____________ 2 ____________

1 ____________ 2 ____________

Main journal 5	1 ____________	2 ____________
	1 ____________	2 ____________
Main journal 6	1 ____________	2 ____________
	1 ____________	2 ____________
Main journal 7	1 ____________	2 ____________
	1 ____________	2 ____________
Rod journal 1	1 ____________	2 ____________
	1 ____________	2 ____________
Rod journal 2	1 ____________	2 ____________
	1 ____________	2 ____________
Rod journal 3	1 ____________	2 ____________
	1 ____________	2 ____________
Rod journal 4	1 ____________	2 ____________
	1 ____________	2 ____________
Rod journal 5	1 ____________	2 ____________
	1 ____________	2 ____________
Rod journal 6	1 ____________	2 ____________
	1 ____________	2 ____________
Rod journal 7	1 ____________	2 ____________
	1 ____________	2 ____________
Rod journal 8	1 ____________	2 ____________
	1 ____________	2 ____________

Tasks completed ____________

3. Journal taper and bell wear measurement requires micrometer measurements to be made in three longitudinal locations and 90 degrees offset from each longitudinal location.

Main journal 1	1 ________	2 ________	3 ________
	1 ________	2 ________	3 ________
Main journal 2	1 ________	2 ________	3 ________
	1 ________	2 ________	3 ________
Main journal 3	1 ________	2 ________	3 ________
	1 ________	2 ________	3 ________
Main journal 4	1 ________	2 ________	3 ________
	1 ________	2 ________	3 ________
Main journal 5	1 ________	2 ________	3 ________
	1 ________	2 ________	3 ________
Main journal 6	1 ________	2 ________	3 ________
	1 ________	2 ________	3 ________
Main journal 7	1 ________	2 ________	3 ________
	1 ________	2 ________	3 ________
Rod journal 1	1 ________	2 ________	3 ________
	1 ________	2 ________	3 ________
Rod journal 2	1 ________	2 ________	3 ________
	1 ________	2 ________	3 ________
Rod journal 3	1 ________	2 ________	3 ________
	1 ________	2 ________	3 ________
Rod journal 4	1 ________	2 ________	3 ________
	1 ________	2 ________	3 ________
Rod journal 5	1 ________	2 ________	3 ________
	1 ________	2 ________	3 ________

Rod journal 6	1 ______	2 ______	3 ______
	1 ______	2 ______	3 ______
Rod journal 7	1 ______	2 ______	3 ______
	1 ______	2 ______	3 ______
Rod journal 8	1 ______	2 ______	3 ______
	1 ______	2 ______	3 ______

Tasks completed ______

Instructor Check/Comments

______

______

______

JOB SHEET 4.2

Name ______________________________ Date ______________

Job Description: Magnetic-flux–testing a crankshaft.

Text Reference: Chapter 4.

Performance Objective: After completing this assignment, the student should be able to magnetic-flux–test a crankshaft and understand what is required to interpret the results.

Protective Clothing: Required for shop exercise sessions: hearing, eye, hand, and foot protection, and shop coat/coveralls.

Tools and Materials:

Crankshaft from a six- or eight-cylinder diesel engine
Magnetic flux test apparatus
OEM service specifications

PROCEDURE

Magnetic-Flux–Test Crankshaft

1. Ensure that the crankshaft is completely clean and that there is no trace of oil remaining in the oil distribution circuit. Mount the crankshaft in the magnetic-flux–test jig.

   Task completed ______________

2. Energize the crankshaft and apply solution. Use the black light (ultraviolet) to identify cracks. Use the OEM service literature to identify the type of crack that requires the crankshaft to be replaced.

   Task completed ______________

3. Identify the location of cracks using the guidelines in the textbook. Evaluate the cracks to determine if they are in critical locations.

   Task completed ______________

Caution: Most small cracks detected by magnetic-flux–testing are harmless.

Task completed ______________

Instructor Check/Comments

____

____

____

JOB SHEET 4.3

Name ______________________________ Date ______________

Job Description: Inspect a connecting rod.

Performance Objective: After completing this assignment, the student should be able to assess the serviceability of a diesel engine connecting rod.

Text Reference: Chapter 4.

Protective Clothing: Required for shop exercise sessions: hearing, eye, hand, and foot protection, and shop coat/coveralls.

Tools and Materials:

Inside micrometer
Outside micrometer and telescoping gauges
Torque wrench
Magnetic flux test equipment
OEM service specifications
Rod test fixture (desirable, but not essential)

PROCEDURE

A rod test fixture is the preferred tool to use when assessing the serviceability of connecting rods, as it speeds up the testing process. However, such fixtures are expensive, and not all educational facilities will have them.

Inspection

1. Clean the connecting rod thoroughly and visually inspect, noting any nicks or cracks anywhere as well as indications of heat discoloration in either of the bearing seats. Evidence of either at this stage should reject the rod.
2. Most OEMs prefer that the rod cap fasteners be replaced at each overhaul. If the old ones are to be reused, visually inspect for any kind of thread or other deformity—NEVER take a risk with a suspect connecting rod cap fastener. Install rod cap (without bearings) and torque to specification.
3. Measure the big end bore for out-of-roundness and taper.
4. Measure the small end bore for out-of-roundness and taper.
5. If a rod test fixture is used, install the conrod assembly and alignment test for twist. Next, test for stretch. This checks the big-end-to-small-end center dimension.
6. Finally, magnetic-flux–test the rod and use the OEM guidelines on what is acceptable in terms of serviceability. NEVER take a risk with a suspect connecting rod.

Tasks completed ______________

Reconditioning

Most OEMs feel that connecting rods should be inspected as outlined previously and replaced if found to be defective. Reconditioning is practiced, however, and conditions such as rod twist and stretch can be repaired. The connecting rod is not to be heated under any circumstances. Big- and small- end eccentricity can be machined concentric when oversize bearings are available. The bottom line on reconditioning is that it is seldom justified economically over the long term.

Instructor Check/Comments

____

____

____

JOB SHEET 4.4

Name ______________________ Date ____________

Job Description: Install a piston kit into a cylinder liner.

Performance Objective: After completing this assignment, the student should be able to assemble a piston pack and install it into the engine cylinder using the correct OEM procedure.

Text Reference: Chapter 4.

Protective Clothing: Required for shop exercise sessions: hearing, eye, hand, and foot protection, and shop coat/coveralls.

Tools and Materials:

- Piston assembly (any type)
- Ring kit
- Connecting rod assembly
- Engine cylinder block with crankshaft installed
- Micrometer kit
- Thickness (feeler) gauges
- Dial bore gauge
- Oil cooling jet alignment tool (Perspex plate and rod)
- Piston ring installation tool
- Piston ring compressor tool (preferably taper type)
- Torque wrench
- Engine (or assembly) lube
- OEM service specifications

PROCEDURE

This exercise can be performed on either a functioning or downed diesel engine. Part of the objective is to correctly identify the components and adhere to the OEM service procedure.

Identify Piston (circle appropriate)

Aluminum trunk   Crosshead   Articulating (Ferrotherm)

Steel trunk (Monotherm)   Steel/comp. trunk (Monocomp)

Identify Rod (circle appropriate)

Square big end   Offset big end

Machined cap   Cracked cap

Tasks completed ____________

1. Mike the piston to specification, crossways at top and bottom of ring belt.

   Task completed ____________

2. Check ring clearances in piston ring grooves using thickness gauges.

   Task completed ____________

3. Measure piston pin boss to spec.

   Task completed ____________

4. Mike wrist pin to spec.

Task completed ________________

5. Measure liner bore to spec using dial bore gauge. Use oil cooling jet alignment tool to check targeting of jet.

Task completed ________________

6. Install each compression ring into the liner and measure ring gap to spec.

Task completed ________________

7. Assemble the piston to the connecting rod. Make sure that the pin is lubed with engine oil; if snap ring retainers are used, the joint should face down. Observe orientation (front/back) arrows!

Task completed ________________

8. Use a ring expander to install each ring into its ring groove. Identify each ring and make sure it is installed with its appropriate side up.

Task completed ________________

9. Install the upper big end friction bearing.

Task completed ________________

10. Generously lube the piston assembly.

Task completed ________________

11. Set ring stagger.

Task completed ________________

12. Ease the piston assembly into a tapered ring compressor. Make sure that the piston assembly does not cock off horizontally as you do this and that the assembly is installed far enough into the ring compressor that all of the rings are radially loaded.

Task completed ________________

13. Generously lube the liner bore with engine oil.

Task completed ________________

14. Maintaining the appropriate orientation (front/back arrows) with the engine, support the ring compressor with both hands and carefully lower the rod into the cylinder bore from above. Take great care not to make contact with rod big end and oil cooling jets.

    Caution! Also take care not to clunk the crank throw.

Task completed ________________

15. Next, get below the cylinder block. Make sure the rod is correctly positioned for pulling into position on the crank throw. Depending on the engine, you may now be able to pull the piston assembly through the ring compressor from below and position the big end. However, in most cases you will have to do this from above the engine.

Task completed ________________

16. Install the lower big end bearing shell. It should be clean and dry on its reverse side. When it is snapped into place in the cap, generously lube the face of the shell.

Task completed ________________

17. Assemble the big end cap to the rod. Keep the rod cap square while seating the cap faces.

Task completed ________________

18. Torque big end caps in increments. Some require final template torque; torque to yield. If the big end cap uses machined faces, check rod sideplay after final torque. Where fractured (cracked) rods are used, checking sideplay is not required.

Task completed ________________

Instructor Check/Comments

____

____

____

INTERNET TASKS

Most engine OEMs use components manufactured and sometimes engineered by subcontractors. Check out some of the following suppliers to engine OEMs and determine what it is they make and for whom:

1. Mahle®
2. Perfect Circle®
3. Delphi®
4. Eaton®

STUDY TIPS

Identify five key points in Chapter 4.

Key point 1 ____________________________

Key point 2 ____________________________

Key point 3 ____________________________

Key point 4 ____________________________

Key point 5 ____________________________

CHAPTER 5

Timing Geartrains, Camshafts, Tappets, Rockers, and Cylinder Valves

OBJECTIVES

After studying this chapter, you should be able to:

- Identify the engine feedback assembly components.
- Identify the engine timing geartrain components.
- Outline the procedure required to time an engine geartrain.
- Define the role of the camshaft in a typical diesel engine.
- Interpret camshaft terminology.
- Perform a camshaft inspection.
- Identify the role valve train components play in running an engine.
- List the types of tappet/cam followers used in diesel engines.
- Inspect a set of push tubes or rods.
- Describe the function of rockers.
- Define the role played by cylinder head valves.
- Interpret valve terminology.
- Outline the procedure required to recondition cylinder head valves.
- Describe how valve rotators operate.
- Perform a valve lash adjustment.
- Outline the consequences of either too much or too little valve lash.
- Create a valve polar diagram.

END OF CHAPTER REVIEW QUESTIONS

1. The location on a cam profile that is exactly opposite the nose is referred to as the
   a. sole.
   b. toe.
   c. ramp.
   d. heel.
2. On a cam profile described as mostly inner base circle, the profiles between IBC and OBC are known as
   a. cam geometry.
   b. ridges.
   c. ramps.
   d. heels.
3. In which direction must a camshaft be rotated on a diesel engine with a crankshaft that is rotated clockwise?
   a. Clockwise (CW)
   b. Counterclockwise (CCW)
   c. Either CW or CCW, depending on the engine
4. Which tool would be required to measure the cam lift of a cylinder block-mounted camshaft in position?
   a. Dial indicator
   b. Outside micrometer
   c. Depth micrometer
   d. Thickness gauges
5. A camshaft gear is usually precisely positioned on the camshaft using a(n)
   a. dial indicator.
   b. interference fit.
   c. key and keyway.
   d. captured thrust washer.
6. When grinding a valve face to remove pitting, the critical specification to monitor during machining would be the
   a. shank diameter.
   b. stem.
   c. poppet diameter.
   d. margin.
7. Which of the following engine running conditions would be most likely to cause a valve float condition?
   a. Engine lug down
   b. Engine overspeed
   c. Operating in the torque rise profile
   d. Operating in the droop curve
8. When grinding a new set of cylinder valve seats, which of the following should be performed before machining?
   a. Install the valves.
   b. Install the rockers.
   c. Adjust the valve yokes.
   d. Install the valve guides.
9. Which of the following methods is used to transfer drive torque from the crankshaft to the camshaft on most commercial diesel engines?
   a. Gears
   b. Belt and pulley
   c. Fluid coupling
   d. Timing chain and sprocket
10. What type of cam geometry is required if the objective is to load the train that rides it for most of the cycle?
    a. Mostly OBC profile
    b. Mostly IBC profile
    c. Symmetrical OBC and IBC profiles

JOB SHEET 5.1

Name ______________________________ Date ______________

Job Description: Adjust valve lash.

Performance Objective: After completing this assignment, the student should be able to adjust the cylinder valves on a diesel engine.

Text Reference: Chapter 5.

Protective Clothing: Required for shop exercise sessions: hearing, eye, hand, and foot protection, and shop coat/coveralls.

Tools and Materials:

Thickness gauges
Torque wrench
Engine barring tool
Chalk
Basic hand tools

PROCEDURE

Most current highway diesel engines use rocker housing cover gaskets, which are reusable provided they are installed clean and are torqued correctly. Depending on the engine, it may not be necessary to reset the bridge adjustment at each routine valve lash adjustment; in any case, adjusting valve lash is a procedure that is more about checking the set dimension than adjusting it.

The following procedure will assume the engine is an inline six-cylinder with a cylinder firing sequence of 1-5-3-6-2-4, of a cylinder block-located camshaft design with no unusual characteristics. Locate the OEM specifications. In most cases, the valve adjustment should be performed cold. In some older engines, this adjustment had to be performed with the engine at running temperature. Make sure that the engine is at the correct temperature for the lash specification; in general, hot lash specs (when available) are lower than cold lash specs. Most engine OEMs today only publish cold lash specs meaning that the engine temperature should be 100°F (38°C) or less during the procedure. This procedure is outlined with the assumption that there is no engine compression brake on the engine.

Check Valve Lash

1. Clean the upper engine with a pressure washer.

   Task completed ______________

2. Remove the rocker housing covers. Locate the engine valve timing indices and determine exactly what position the engine is in by observing the valves using the companion cylinder method. The valve adjustment can begin at any engine cylinder provided the correct sequence is observed.

   Task completed ______________

3. With the engine located at the timing indices for the cylinder to be adjusted, check the valve clearance by inserting the correct size thickness gauge between the rocker pallet and the valve stem or bridge. Minimal drag is required. If the setting is correct, chalk mark the rocker and bar the engine 120 degrees to the next valve to be set.

   Task completed ______________

Adjust Valve Lash

1. If the valve lash is either loose or tight, the lash will have to be correctly set. Back off the locknut on the rocker adjusting screw. Initially back off the adjusting screw.

   Task completed ________________

2. Screw down the adjusting screw until the rocker pallet starts to load down the valve or bridge, then back off the adjusting screw sufficiently to insert the correct size thickness gauge. Release the thickness gauge—this is important. Do not attempt to set the valve lash dimension while simultaneously moving the thickness gauge. Screw the adjusting screw CW until resistance is felt, then turn the adjusting screw an additional half a flat/one flat. Continue to hold the adjusting screw with a screwdriver and use a torque wrench and crowsfoot socket to torque the locknut.

   Task completed ________________

3. Check the thickness gauge drag. It should be minimal. Despite what you may be told, cylinder valves should not be set tight; they should be set to the OEM-specified dimension.

   Task completed ________________

4. Continue to set the valves in the engine firing sequence. Run the engine after the adjustment. A valve with excessive lash can usually be heard. Remember, at cold startup, correctly set valves will usually produce more clatter than a set of valves set tight over specification. The idea of valve adjustment is to produce the correct performance at operating temperatures.

   Task completed ________________

Instructor Check/Comments

____

____

____

JOB SHEET 5.2

Name ______________________________ Date ______________

Job Description: Map a valve polar diagram.

Performance Objective: Using a four-stroke cycle diesel engine, map an accurate valve polar diagram to identify valve status during each phase of the four-stroke cycle.

Text Reference: Chapter 5.

Protective Clothing: Required for shop exercise sessions: hearing, eye, hand, and foot protection, and shop coat/coveralls.

Tools and Materials:

Assembled diesel engine
Engine barring tools
Protractor and chalk

PROCEDURE

1. Identify the engine you will be working on. Note the firing order.

   Task completed ______________

2. Identify the specific cylinder from which you will draw data; common sense dictates that this should be either #1 (front) or #6 (rear) in an inline six-cylinder engine. For the purpose of this exercise we will assume you are working on the #1 cylinder on a six-cylinder engine.

   Task completed ______________

3. Begin with #1 piston at TDC completing its compression stroke. Its companion cylinder, #6, should have its valves rocking at overlap. Both valves over #1 cylinder will be closed. Check for valve lash on both intake and exhaust valves.

   Task completed ______________

4. Bar engine in its correct direction of rotation. This moves the piston on #1 downward (power stroke). Note on the protractor the exact point when the exhaust valve(s) begin to open: ______________ degrees ATDC.

   Task completed ______________

5. Continue to bar the engine in its correct direction of rotation. This will take #1 piston through BDC and upward through the exhaust stroke toward TDC.

   Task completed ______________

6. Beginning of valve overlap. Note on the protractor the exact point at which the intake valve(s) open: ______________ degrees BTDC.

   Task completed ______________

7. End of valve overlap. Observing the exhaust valve(s), note on the protractor the exact point at which they close to end overlap: ______________ degrees ATDC.

   Task completed ______________

8. Intake stroke. The piston will now travel downward through the intake stroke and pass through BDC. Note the exact point the intake valve(s) close to begin the compression stroke: ______________ degrees BTDC.

Task completed ______________

9. Using the data you have collected from the above procedure, map the valve status through the four-stroke cycle using a polar diagram such as that in **Figure 5–1**.

Task completed ______________

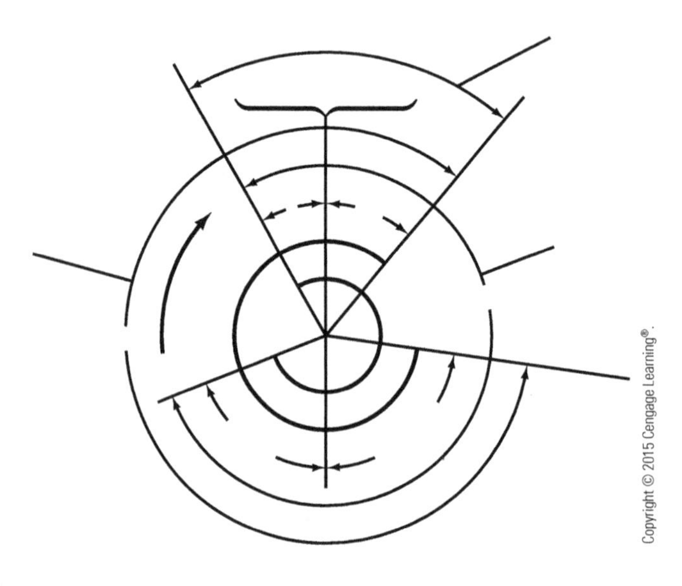

FIGURE 5–1 Valve polar diagram.

Instructor Check/Comments

____

____

____

JOB SHEET 5.3

Name ______________________________ Date ______________

Job Description: Recondition cylinder head valves.

Performance Objective: After completing this assignment, the student should be able to assess the serviceability of a set of diesel engine cylinder valves and machine them as required.

Text Reference: Chapter 5.

Protective Clothing: Required for shop exercise sessions: hearing, eye, hand, and foot protection, and shop coat/coveralls.

Tools and Materials:
- Bead blaster
- Outside micrometer
- Valve refacer
- OEM specifications

PROCEDURE

When servicing cylinder valves, it is good practice to ensure that they remain matched with their location in the cylinder head. A numbered valve template is the best way of ensuring this.

Inspection

1. Clean the valves using a glass bead blaster. Obtain instruction on the safety precautions that must be observed before operating this equipment.

   Task completed ______________

2. Visually inspect each valve, looking for cracks and heat deformation. Pay special attention to the fillet and seat.

   Task completed ______________

3. Measure the valve stem using a micrometer, and compare results to specification.

   Task completed ______________

Reconditioning

1. Determine the valve angle by consulting the OEM specifications. Never assume that an interference angle fit is required—it almost never is in modern engines. Dress the valve grinding stone with a dressing tool.

   Task completed ______________

2. Install the valve securely in the machine chuck and adjust the table stop so that the grinding stone cannot contact the valve stem.

   Task completed ______________

3. Set the machine oil jet so that it is directed to flow over the valve seat.

   Task completed ______________

4. Use the table control lever to move the valve through the reciprocating stroke required to machine it, beginning so that it is just in contact with the stone. Throughout the machining procedure, advance the table slowly so that no cut is too aggressive.

   Task completed ______________

5. Machine the valve until the seat is free of any heat discoloration and pitting. At this point, measure the valve margin. This specification is critical and will determine if the valve can be returned to service; see **Figure 5–2**. Do not risk reusing a valve with a below-specification valve margin because a rapid failure will result.

Task completed ________________

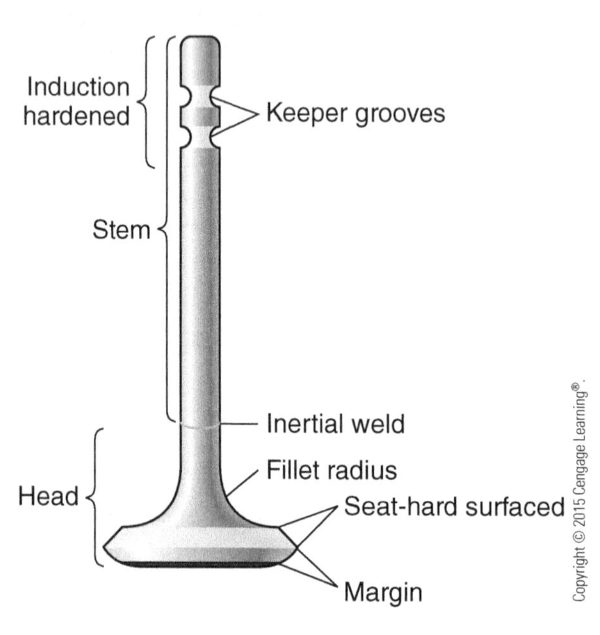

FIGURE 5–2 Valve terminology.

6. Machine the tip of the valve stem. This removes the "mushroom" that can result from long service. Dress the stem so that a small chamfer appears.

Task completed ________________

7. Return the reconditioned valve to its location in the template and repeat the above process with the next valve.

Task completed ________________

Instructor Check/Comments

____

____

____

INTERNET TASKS

Use Internet search engines and other information resources to research the following:

1. Identify two current overhead camshaft HD commercial diesel engines.
2. Identify an engine that uses concept gears and list the advantages.
3. Identify three engines with camshafts that rotate CW and three that rotate CCW.

STUDY TIPS

Identify five key points in Chapter 5.

Key point 1 ____________________________

Key point 2 ____________________________

Key point 3 ____________________________

Key point 4 ____________________________

Key point 5 ____________________________

CHAPTER 6

Cylinder Blocks, Liners, Cylinder Heads, Rocker Housings, Oil Pans, and Manifolds

OBJECTIVES

After studying this chapter, you should be able to:

- Identify the components classified as engine housing components.
- Identify the types of cylinder blocks used in diesel engines.
- Outline the procedure required to inspect a cylinder block.
- Measure an engine block to specifications using service literature.
- Identify the types of cylinder liners used in diesel engines.
- Explain the procedure required to remove dry, wet, and midstop liners.
- Perform selective fitting of a set of dry liners to a cylinder block.
- Explain how cavitation erosion occurs on wet liners.
- Identify the types of cylinder heads used in diesel engines.
- Describe the component parts of a cylinder head.
- Define *component creep* and *gasket yield*.
- Explain the procedure required to measure, test, and recondition a cylinder head.
- Describe the role of the intake and exhaust manifolds.
- Describe the function of the oil pan in the engine.

END OF CHAPTER REVIEW QUESTIONS

1. Which of the following cylinder block designs would be the most common in commercial diesel engines?
   a. 8-cylinder, 90-degree, V-configuration
   b. Inline, 6-cylinder
   c. 6-cylinder, 60-degree, V-configuration
   d. Inline, 8-cylinder
2. The main reason for hydrostatically bench pressure testing a diesel engine cylinder head is to
   a. test cylinder gas leakage.
   b. check for air leaks.
   c. check for exhaust leaks.
   d. check for coolant leaks.
3. When selective fitting a set of dry liners to a cylinder block, which of the following statements should be true?
   a. The liner with the largest OD is fitted to the bore with the largest ID.
   b. The liners are installed with an interference fit.
   c. The liner with the smallest OD is fitted to the bore with the largest ID.
   d. The liners are installed with a fractionally loose fit.
4. Which of the following is the recommended method of pulling dry liners from a cylinder block?
   a. A puller and shoe assembly
   b. Heat and shock cool method
   c. Fracture the liner in position
   d. Oxyacetylene torch
5. Which of the following is recommended for checking a cylinder block line bore?
   a. Straightedge and thickness gauges
   b. Dial bore gauge
   c. Dial indicator
   d. Master line bore bar
6. When boiling out a cylinder block, why is it critical that all the scale in the water jacket be removed?
   a. Scale is an effective insulator
   b. Scale may contaminate the engine lubricant
   c. Scale can accelerate coolant silicate drop-out
   d. Scale causes cavitation of wet liners
7. Which of the following tools should be used to check a cylinder block deck for warpage?
   a. Master bar
   b. Dial indicator
   c. Laser
   d. Straightedge and feeler gauges
8. Which of the following tools should be used to check a cylinder head for warpage?
   a. Master bar
   b. Dial indicator
   c. Laser
   d. Straightedge and thickness gauges
9. When cavitation damage can be observed on a set of wet liners, which of the following is most likely to be responsible?
   a. Lubrication breakdown
   b. High cylinder pressures
   c. Coolant breakdown
   d. Cold engine operation
10. Once a cylinder head gasket fire ring has been torqued to its yield point, it should
    a. not be reused after removal.
    b. be immediately heated to operating temperature.
    c. deform and no longer seal effectively.
    d. be pressure tested with shop air.

JOB SHEET 6.1

Name ______________________________ Date ______________

Job Description: Replace a cylinder head gasket.

Performance Objective: After completing this assignment, the student should be able to remove a cylinder head, clean and inspect the cylinder head and cylinder block decks, replace the cylinder head gasket, and reassemble the engine. The following procedure assumes that the cylinder head gasket has been diagnosed as having failed and that the cylinder head itself does not require servicing.

Text Reference: Chapter 6.

Protective Clothing: Required for shop exercise sessions: hearing, eye, hand, and foot protection, and shop coat/coveralls.

Tools and Materials:

Assembled diesel engine
Thickness gauges
Straightedge
Injector removal tools
Lifting T bolts
Cherry picker hoist
Depth micrometer
New head gasket
New manifold gaskets
Air tools
Basic hand tools

PROCEDURE

Cylinder Head Removal

1. If the engine is in-chassis, pressure-wash the engine and engine compartment.

   Task completed ______________

2. Drain coolant from the engine. Store away from the work area in a sealed container.

   Task completed ______________

3. Consult OEM procedure for removal of cylinder head. In most cases, it is permissible to use air tools throughout the removal process.

   Task completed ______________

4. It is strongly advised that the injector assemblies be removed from the cylinder head before removal; the nozzle tips usually protrude from the cylinder head and can easily be damaged when installed.

   Task completed ______________

5. When lifting the cylinder head assembly from the block, make sure that the weight of the cylinder head assembly is accounted for. Multisection cylinder heads can often easily be removed by hand using lifting Ts. However, single-slab cylinder head castings on inline six-cylinder engines should be removed using a hoist.

   Task completed ______________

Cleaning and Inspection

1. Use a scraper and then fine-grade emery cloth to thoroughly clean both the cylinder head and the cylinder block deck. Visually inspect both the cylinder head and block.

   Task completed ________________

2. Use an air blower to clean out coolant, lubricant, and bolt passages and holes.

   Task completed ________________

3. Check the cylinder head for warpage using a straightedge and thickness gauges.

   Task completed ________________

4. Check the cylinder block deck for warpage using a straightedge and thickness gauges. If steps 3 and 4 do not meet the OEM-required specification, further servicing not outlined in this task sheet is required.

   Task completed ________________

5. Check cylinder liner protrusion using a depth micrometer to make sure that this meets specification.

   Task completed ________________

Reassembly

Make sure that both the cylinder head and cylinder block deck are clean and dry.

1. Install the cylinder head gasket onto the block deck. If the fire rings and lubricant and coolant grommets are not integral with the gasket, carefully place them in position. If visibility is restricted or the cylinder block deck is not horizontal (V-configured engine), consider using dowels to ensure the head is properly aligned with the deck. In most cases, the cylinder head gasket should be installed completely dry; unless recommended by the OEM, sealants should never be used.

   Task completed ________________

2. Lightly lubricate the cylinder bolt threads and the surface area under the fastener cap, then insert it in the bolt holes. Locate the OEM torque specifications. Sequentially and incrementally torque the fasteners to specification. **Figure 6–1** shows the torque sequence favored by one OEM. Note how the sequence runs from the center outward, which is common in almost all single-slab cylinder heads.

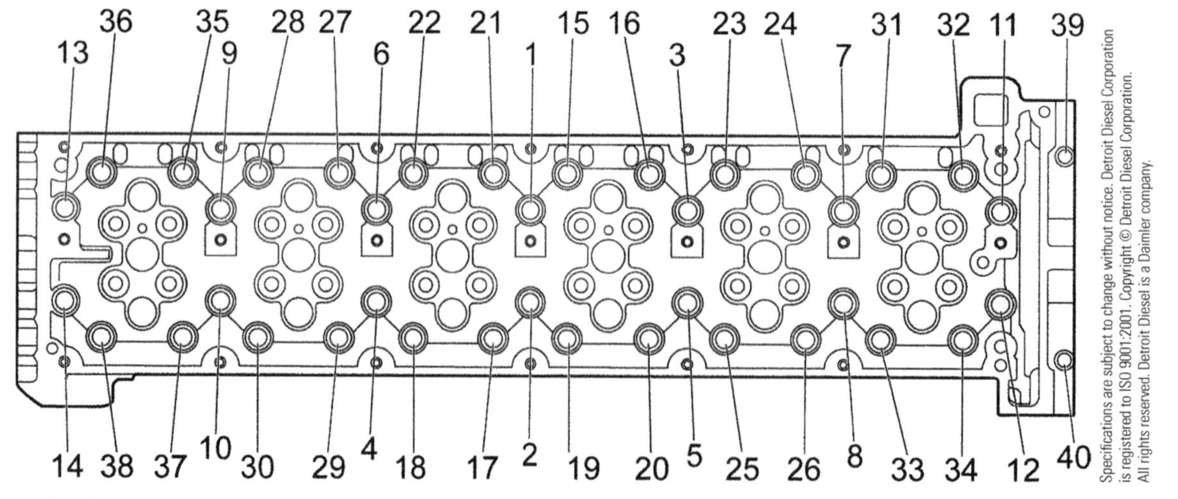

FIGURE 6–1 Torque sequence.

   Task completed ________________

3. Perform an engine overhead adjustment as outlined by the OEM. At minimum, this requires setting valve lash; in most engines, this will usually require setting and timing the injector assemblies.

Task completed ________________

4. Reassemble engine components and change or replace fluids as outlined by the OEM.

Task completed ________________

5. Performance test the engine and check for leaks.

Task completed ________________

Instructor Check/Comments

____

____

____

JOB SHEET 6.2

Name ____ Date ____________________

Job Description: Set up a dial bore gauge and measure cylinder bore.

Performance Objective: After completing this assignment, the student should be able to interpret OEM cylinder bore specifications, set up a dial bore gauge, and measure cylinder bore.

Text Reference: Chapters 2 and 6.

Protective Clothing: Required for shop exercise sessions: hearing, eye, hand, and foot protection, and shop coat/coveralls.

Tools and Materials:

Disassembled diesel engine. This exercise will work using parent bore, dry liner cylinder bore, dry liner bore, or wet liner bore.
Dial bore gauge
Snap gauge set
OEM service specifications

PROCEDURE

The procedure for setting up a dial bore gauge is described in some detail in Chapter 2. Using a dial bore gauge is the only method of measuring cylinder and liner bores in diesel engines; other methods are simply not as fast or involve indirect measurement techniques.

1. Identify the specification window by referring to the OEM data. See Chapter 2 for information about rounding specs made in units of 0.0001-inch up or down. When you take a look at the OEM specification window for a bore, round to the nearest ½-thousandth and identify the *maximum* bore specification. This is important. It means that when you start making measurements, every acceptable specification will read on the *minus* side of the dial bore gauge zero. Use **Figure 6–2** to record your data.

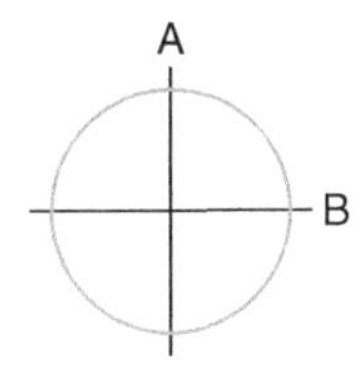

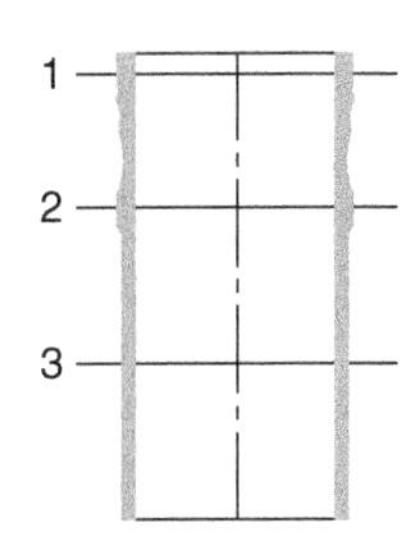

1 A1 ________	B1 ________		2 A1 ________	B1 ________		3 A1 ________	B1 ________
A2 ________	B2 ________		A2 ________	B2 ________		A2 ________	B2 ________
A3 ________	B3 ________		A3 ________	B3 ________		A3 ________	B3 ________
4 A1 ________	B1 ________		5 A1 ________	B1 ________		6 A1 ________	B1 ________
A2 ________	B2 ________		A2 ________	B2 ________		A2 ________	B2 ________
A3 ________	B3 ________		A3 ________	B3 ________		A3 ________	B3 ________

Maximum specification ________ (set micrometer to this dimension)

Minimum specification ________

FIGURE 6–2 Cylinder bore data recording chart.

Task completed ____________

2. Setting the micrometer: Set the micrometer reading to the *maximum* permitted bore specification, then lock the micrometer. Record the maximum specification on the bore data chart shown in Figure 6–2. Next, wrap the micrometer frame in a wiper and gently but firmly clamp it into a vise.

   Task completed ______________

3. Setting the dial bore gauge to the mike dimension. Select the appropriate length adjustable guide (for the dial bore gauge) and screw inward into the sled until the dimension between the adjustable guide and the measuring plunger is less than the micrometer setting. Supporting the measuring plunger against the micrometer anvil, begin to screw the adjustable guide counterclockwise (CCW) until it makes contact with the micrometer spindle. Now as you continue to rotate the adjustable guide CCW, the indicator reading needle will begin to rotate; make sure that it rotates at least one full revolution (on a typical indicator, this will be 100 thousandths). Remove the dial bore gauge. Lock the jam nut on the adjustable guide. It does not matter if the adjustable guide moves slightly as you are engaging the jam nut. Next, insert the dial bore gauge back into the mike. Make sure the dial bore indicator turns through approximately one rotation of travel, then zero the indicator and lock the setting. Having set the dial bore gauge, remove and reinstall it a couple of times to check your measurement. Each time you install the dial bore gauge into the mike it should read exactly zero. When it does, the zero corresponds to the *maximum* permitted bore specification. **This means that *any* positive reading on the indicator is out of spec.** It also means that when you record your data, every in-spec reading should be a minus reading in a within-spec bore.

   Task completed ______________

4. When you are making bore measurements to determine the serviceability of cylinder liners, they should be made in the following locations as shown in the figure that accompanies this job sheet: (1) top of the ring belt sweep; (2) bottom of the ring belt sweep; and (3) midway between measurement 2 and the bottom of the liner. To make a bore measurement, gently hold the dial bore gauge between two fingers on the grip of the handle above the indicator. Allow the measuring sled to pivot in the bore, using the dial bore handle to sweep the device through an arc. Watch the needle as you move the dial bore gauge through each sweep. You are looking for the stroke-over point, which is where you record your spec.

   Task completed ______________

5. In recording the measurements, you want to keep things as simple as possible. Think in terms of relative thousandths. Avoid recording the full specification; just work in units of thousandths (0.001-inch) and half thousandths. Complete the chart for each cylinder you measure.

   Task completed ______________

6. Assessing the results: If you have used the method outlined in this book, every measurement should be a negative one if the bore is within spec. The beauty of this method is that any positive measurement is outside the required specification, so you can easily identify it.

   Task completed ______________

Instructor Check/Comments

____

____

____

INTERNET TASKS

See what you can find out about the latest material used in diesel engine cylinder blocks and how some OEMs use parent bore but treat the bore with induction hardening. Look for information on how cylinder blocks are machined and the procedures different OEMs recommend for cylinder head servicing. Type the following key words into your search engine and see what you come up with.

- Compacted graphite iron (CGI)
- Cummins cylinder blocks
- Line bore machining
- Magnetic-flux–testing services

STUDY TIPS

Identify five key points in Chapter 6.

Key point 1 ______________________________

Key point 2 ______________________________

Key point 3 ______________________________

Key point 4 ______________________________

Key point 5 ______________________________

CHAPTER 7

Engine Lubrication Systems

OBJECTIVES

After studying this chapter, you should be able to:

- Identify the main components of a typical diesel engine lubrication circuit.
- List the properties of heavy-duty engine oils.
- Define the term *hydrodynamic suspension.*
- Describe the difference between thin film and thick film lubrication.
- Interpret the terminology used to classify lubrication oil.
- Interpret API classifications and SAE viscosity grades.
- Replace and properly calibrate a lube oil dipstick.
- Describe the two types of oil pumps commonly used on diesel engines.
- Describe the operation of an oil pressure regulating valve.
- Define the term *positive filtration.*
- Outline the differences between *full flow* and *bypass* filters.
- Service a set of oil filters.
- Outline the role of an oil cooler in the lubrication circuit.
- Test an oil cooler core using vacuum or pressure testing.
- Identify the methods used to signal oil pressure in current diesel engines.
- Outline the procedure for taking an engine oil sample for analysis.
- Interpret the results of a laboratory oil analysis.

END OF CHAPTER REVIEW QUESTIONS

1. Which of the following SAE multigrade oils is most likely to be used by commercial fleets during North American summertime conditions?
    a. 5W-30
    b. 15W-40
    c. 20W-20
    d. 20W-50

2. Which of the following API classifications would indicate that the oil was formulated for diesel engines meeting 2007 emission standards and later engine model years?
   a. SF
   b. CC
   c. CI-4
   d. CJ-4
3. Which of the following conditions could result from a high crankcase oil level?
   a. Lube oil aeration
   b. Oil pressure gauge fluctuations
   c. Friction bearing damage
   d. All of the above
4. A full flow oil filter has become completely plugged. Which of the following is likely if the engine is running?
   a. A bypass valve diverts the oil around the filter
   b. Engine lubrication ceases
   c. Oil pump hydraulically locks
   d. Engine seizure occurs
5. Which type of oil pump is most commonly used in current highway diesel engines?
   a. External gear
   b. Plunger
   c. Centrifugal
   d. Vane
6. Which of the following is usually the preferred OEM method of testing an oil cooler bundle?
   a. Vacuum test
   b. Die penetrant test
   c. Hydrostatic test
   d. Pressure test using shop air and a bucket of water
7. When interpreting a used engine oil analysis profile, which of the following conditions would be most likely to cause high silicon levels?
   a. Plugged air cleaner
   b. A hole in the air cleaner
   c. Turbocharger bearing failure
   d. Oil cooler bundle disintegration
8. When oil has a milky, clouded appearance, it is probably contaminated with
   a. fuel.
   b. dust.
   c. engine coolant.
   d. air.
9. Which of the following statements correctly describes viscosity?
   a. Lubricity
   b. Resistance to heat
   c. Resistance to shear
   d. Breakdown resistance
10. Technician A states that 15W-40 is the most commonly used truck engine oil. Technician B states that 15W-40 oil is not an appropriate all-season oil because winter temperatures are often much lower than the SAE recommended ambient temperature window for this oil. Who is correct?
    a. Technician A only
    b. Technician B only
    c. Both A and B
    d. Neither A nor B

JOB SHEET 7.1

Name ______________________________ Date ______________

Job Description: Service a diesel engine.

Performance Objective: After completing this assignment, the student should be able to change the engine oil, oil filters, fuel filters, and coolant filter.

Text Reference: Chapter 7.

Protective Clothing: Required for shop exercise sessions: hearing, eye, hand, and foot protection, and shop coat/coveralls.

Tools and Materials:

Functional highway diesel engine
Sufficient quantity of engine oil to replace oil in the sump
OEM diagnostic software and online service information system (SIS)

PROCEDURE

The procedure for changing engine oil varies depending on the actual equipment you are working on. You should log the service procedure in the vehicle service file by accessing the chassis data bus.

1. Park the vehicle on a level floor and chock the wheels. Place a drain barrel under the oil pan. Using caution (engine oil can be hot), remove the drain plug and dump oil.

Task completed ______________

2. Take an oil sample midway through the flow of oil exiting the crankcase.

Task completed ______________

3. Remove the oil full flow filters and seal for disposal. **Figure 7–1** shows the filter arrangement used on a Volvo VDE-16 engine.

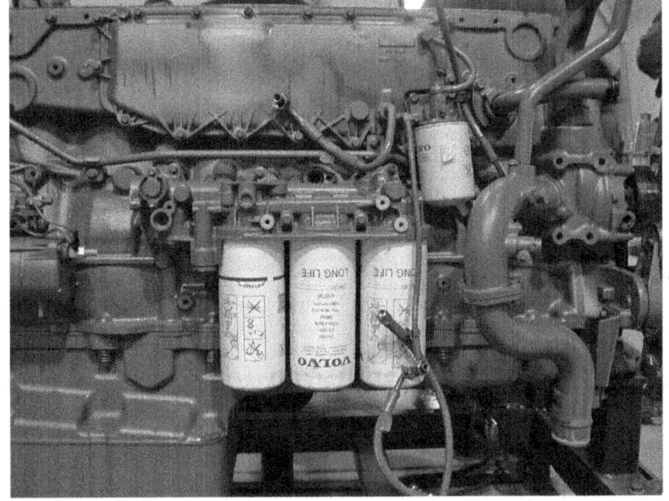

FIGURE 7–1 Volvo VDE-16 filter pad.

Task completed ______________

4. Remove the bypass oil filters and seal for disposal.

Task completed ________________

5. Install new full flow filters. Prime each filter with new engine oil.

Task completed ________________

6. Install new bypass oil filters. Check method of purging bypass canister(s).

Task completed ________________

7. Remove and seal the fuel filters for disposal.

Task completed ________________

8. Install new fuel filters. Direct-prime the primary filter. Install secondary filter dry and then hand prime (or electric prime if system is so equipped). You should crack an exit line from the secondary filter pad so you can observe when circuit is properly primed.

Task completed ________________

9. Remove coolant filter. Test coolant chemistry using the appropriate OEM test kit. Install a new coolant filter dry.

Task completed ________________

10. Check air filter restriction gauge. If within maximum spec and under 12 months of service life, leave it in place. If over in either category, replace the air cleaner.

Task completed ________________

11. Install and torque the oil pan plug using a new sealing washer. *Check torque again.*

Task completed ________________

12. Install the appropriate quantity of new engine oil in the crankcase.

Task completed ________________

13. Check crankcase oil level.

Task completed ________________

14. Start engine. Run at idle for one minute while checking for leaks.

Task completed ________________

15. Leave for one minute then check crankcase oil level with dipstick and add oil if required.

Task completed ________________

16. Check oil pan plug torque one more time.

Task completed ________________

17. Run engine for five minutes, checking for leaks.

Task completed ________________

18. Make sure that no residues of the service you have performed on the engine are visible. Clean with a high pressure washer if necessary.

Task completed ________________

Instructor Check/Comments

____

____

____

INTERNET TASKS

Check out some of the specialty suppliers of diesel engine oils and filters you might know. Investigate the viability of using synthetic oil in a diesel engine. Enter the following key words into your search engine and see what you come up with.

1. Valvoline diesel engine oils
2. Shell Rotella T diesel engine oils
3. Lubrizol
4. Quaker State diesel engine oils
5. Fleetguard
6. Pennzoil® diesel engine oils
7. Texaco diesel engine oils

STUDY TIPS

Identify five key points in Chapter 7.

Key point 1 ____________________________

Key point 2 ____________________________

Key point 3 ____________________________

Key point 4 ____________________________

Key point 5 ____________________________

CHAPTER 8

Engine Cooling Systems

OBJECTIVES

After studying this chapter, you should be able to:

- Identify diesel engine cooling system components and their principles of operation.
- Define the terms *conduction*, *convection*, and *radiation*.
- Identify the three types of coolant used in current highway diesel engines.
- Outline the properties of a heavy-duty antifreeze.
- Calculate the boiling and freeze points of a coolant mixture.
- Mix coolant using the correct proportions of water, antifreeze, and SCAs.
- Perform standard SCA tests and measure antifreeze protection.
- Identify the problems scale buildup can create in an engine cooling system.
- List the advantages claimed for extended life coolants.
- Outline the causes of wet liner cavitation and the steps required to minimize it.
- Identify the types of heavy-duty radiators, including *downflow, crossflow*, and *counterflow*.
- Test a radiator for external leakage using a standard cooling system pressure tester.
- Test a radiator cap.
- Identify the different types of thermostats in use and describe their principle of operation.
- Describe the role of the coolant pump.
- Define the role of the coolant filters and their servicing requirements.
- List the types of temperature gauges used in highway diesel engines.
- Describe how a coolant level warning indicator operates.
- Define the roles played by the shutters and engine fan in managing engine temperatures.
- Outline the operation of an actively pressurized cooling system (APCS).
- Diagnose basic cooling system malfunctions.

END OF CHAPTER REVIEW QUESTIONS

1. Which type of diesel engine coolant is regarded as potentially the most harmful when considered from a maintenance and handling point of view?
   a. Ethylene glycol
   b. Propylene glycol
   c. Pure water
   d. Extended life coolant
2. What causes wet liner cavitation?
   a. Aerated coolant
   b. Combustion gas leakage
   c. Air in the radiator
   d. Vapor bubble collapse
3. What causes cooling system hoses on an engine to collapse when the unit is left parked overnight?
   a. This is normal
   b. Defective thermostat
   c. Improper coolant
   d. Defective radiator cap
4. When a radiator cap pressure valve fails to seal, which of the following would be most likely to occur?
   a. Coolant boil-off
   b. Cooler operating temperatures
   c. Higher HC emissions
   d. Cavitated cylinder liners
5. In a typical diesel-powered highway truck at operating temperature, which of the following should be true?
   a. Coolant temperatures run cooler than lube oil temperatures.
   b. Coolant temperatures run warmer than lube oil temperatures.
   c. Coolant temperatures should be equal to lube oil temperatures.
6. What operating principle is used by a typical diesel engine coolant pump?
   a. Positive displacement
   b. Centrifugal
   c. Constant volume
   d. Opposed gear type
7. When the thermostat routes the coolant through the bypass circuit, what is happening?
   a. The coolant is cycled primarily through the radiator.
   b. The coolant is cycled primarily through the engine.
   c. The coolant is cycled primarily through auxiliary heat exchangers.
8. Fiberglass fans with flexible pitch blades are designed to drive air at greatest efficiency at
   a. low speeds.
   b. all speeds.
   c. high speeds.
9. In the event of an engine overheating, where is the coolant likely to boil first?
   a. Engine water jacket
   b. Top radiator tank
   c. Inlet to the coolant pump
   d. Thermostat housing
10. What instrument do most engine OEMs recommend for checking the degree of antifreeze protection in a heavy-duty diesel engine coolant?
    a. Hydrometer
    b. Refractometer
    c. Spectrographic analyzer
    d. Color-coded test coupon

JOB SHEET 8.1

Name ______________________________ Date ______________

Job Description: Pressure-test a cooling circuit.

Performance Objective: After completing this assignment, the student should be able to perform an in-chassis pressure test on a diesel engine cooling system according to OEM procedure. This test is often performed to locate a cooling circuit external leak.

Text Reference: Chapter 8.

Protective Clothing: Required for shop exercise sessions: hearing, eye, hand, and foot protection, and shop coat/coveralls.

Tools and Materials:

Functional diesel engine
Coolant pressure test kit and adaptors
Hand tools

PROCEDURE

1. Identify the OEM system maximum pressure specification and verify that the correct radiator cap is installed. Many systems also have a pressure relief valve, which is usually set to crack within 2 psi above the specified maximum system pressure.

Task completed ______________

2. Use the correct procedure to remove the radiator cap; make sure that it is not removed from the radiator neck until the system pressure is relieved, or the result can be a serious burn. Locate the correct adaptor and insert it in the radiator neck. Actuate pump to pressurize the cooling system to the specified system pressure value.

Task completed ______________

3. Use a trouble light or focused beam flashlight to check for system leaks. The use of a pocket-sized flashlight capable of beam focus (Mini-Mag™) is preferred.

Task completed ______________

Caution: Cooling system leaks may occur only when the coolant is cold or only when the coolant is at operating temperature. Cold leaks are often associated with loose hose clamps. When attempting to locate a hot leak, exercise caution in removing radiator caps and adaptors during testing.

Instructor Check/Comments

____

____

____

INTERNET TASKS

Use a search engine to research the following terms and download some information on different types of antifreeze. See if you can find out what is recommended for operation in arctic conditions.

- Texaco ELC
- Fleetguard antifreeze
- Propylene glycol
- Ethylene glycol

STUDY TIPS

Identify five key points in Chapter 8.

Key point 1 ____________________________

Key point 2 ____________________________

Key point 3 ____________________________

Key point 4 ____________________________

Key point 5 ____________________________

CHAPTER 9

Engine Breathing

OBJECTIVES

After studying this chapter, you should be able to:

- Identify the intake and exhaust system components.
- Describe how intake air is routed to the engine's cylinders.
- Describe how exhaust gases are routed out to aftertreatment devices.
- Define the term *positive filtration*.
- Outline the operating principle of an air precleaner.
- Service a dry, positive air cleaner.
- Perform an inlet restriction test.
- Identify the subcomponents on a truck diesel engine turbocharger.
- Define constant and variable geometry turbochargers.
- Outline the operating principles of turbochargers.
- Troubleshoot common turbocharger problems.
- Define the role of a charge air cooler in the intake circuit.
- Test a charge air heat exchanger for leaks.
- Relate valve configurations and seat angles to breathing efficiency.
- Outline the role of a diesel engine muffler device.
- Identify the different types of catalytic converters used on current diesels.
- Describe the operation of EGR and DPF systems.

END OF CHAPTER REVIEW QUESTIONS

1. Which tool should be used to accurately check the inlet restriction of a dry air filter?
   a. Mercury manometer
   b. Trouble light
   c. Negative pressure gauge
   d. Restriction gauge
2. Which of the following would be a typical maximum specified inlet restriction for an air filter on a turbocharged diesel?
   a. 25 inches water
   b. 25 inches mercury
   c. 25 psi
   d. 25 kPa

3. Which of the following types of filters has the highest filtering efficiencies?
   a. Centrifugal precleaners
   b. Oil bath
   c. Dry, positive
4. Which of the following should be performed first when checking an engine that produces black smoke under load?
   a. Injection timing
   b. Plugged particulate trap
   c. Fuel chemistry analysis
   d. Air filter restriction test
5. Which of the following best describes the function of a C-EGR system on a diesel engine?
   a. Increases engine breathing efficiency
   b. Dilutes intake charge with cooled dead gas
   c. Preheats the intake charge to the cylinder
   d. Assists the turbocharger in boosting intake pressure
6. Which of the following best describes the function of a wastegate in a current turbo-boosted diesel engine?
   a. Bleeds down intake boost air when excessively high
   b. Options exhaust gas to bypass the turbine
   c. Adjusts the volute flow area
   d. Holds the exhaust valves open when turbo-boost is low boost
7. What is the critical flow area managed by a VG turbo using a variable nozzle operating principle?
   a. Compressor diffuser
   b. Impeller inlet throat
   c. Turbine volute
   d. EGR gate
8. The constant geometry turbocharger used with a high torque rise highway diesel engine is usually designed to produce its best efficiency at which rpm?
   a. Rated
   b. Top engine limit (TEL)
   c. Peak torque
   d. High idle
9. Which of the following components would be located in a diesel exhaust gas aftertreatment canister?
   a. Diesel particulate filter
   b. Catalytic converter(s)
   c. Muffler
   d. Any or all of the above
10. Which of the following best describes the function of a VG turbocharger?
    a. Behaves like a large turbocharger when engine load is high
    b. Behaves like a small turbocharger when engine load is high
    c. Reroutes NO_x back into the engine cylinders
    d. Reroutes EGR gas into the DPF

JOB SHEET 9.1

Name ______________________________ Date ______________

Job Description: Inspect and test some key engine intake and exhaust system components.

Performance Objective: After completing this assignment, the student should be able to inspect and test some key engine intake and exhaust system components using manufacturer-recommended tools and procedures.

Text Reference: Chapter 9.

Protective Clothing: Required for shop exercise sessions: hearing, eye, hand, and foot protection, and shop coat/coveralls.

Tools and Materials:

Operational diesel engine
Restriction gauges
Negative pressure gauges
Manometer

PROCEDURE

The objective of this exercise is to identify some of the main breathing circuit components on a specific engine. To help you to understand what to look for, refer to **Figure 9–1**; note that on some post-2010 engines, an SCR system may also be used.

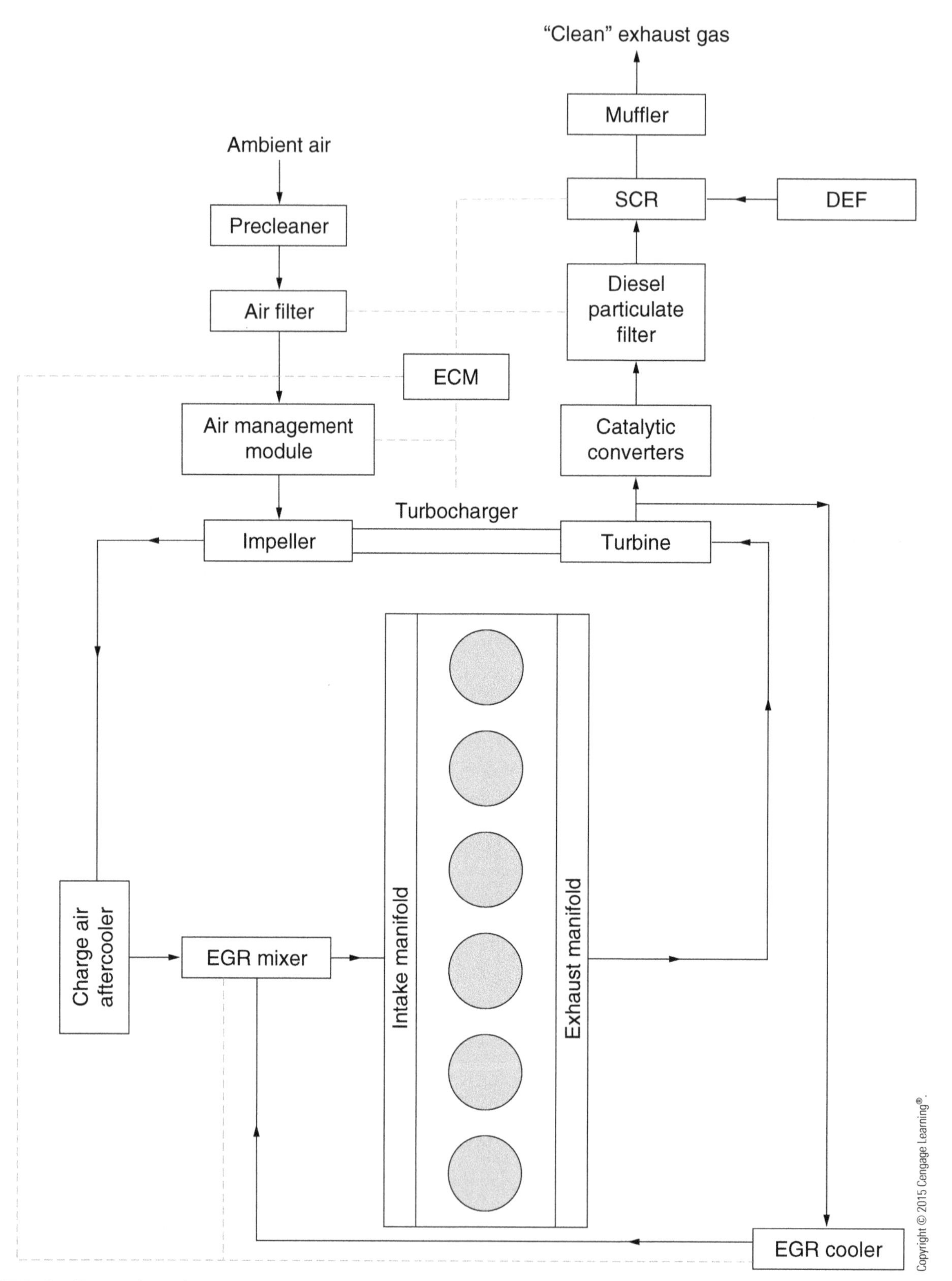

FIGURE 9–1 Engine breathing circuit components.

1. Student to identify:
   - Engine manufacturer ___
   - Engine model or designation ___
   - Engine displacement ___
   - Engine torque rating ___
   - Engine peak power rating ___ (BHP/kW)

Task completed ___

2. Student to identify the filtration system components.
   - Prefilter—yes ___ no ___
   - Main filter type ___
   - Restriction gauge reading ___ H_2O

Task completed ___

3. Remove air filter restriction gauge from port in air cleaner canister. Fit an H_2O manometer or accurate negative pressure gauge. Run engine and perform three throttle snaps. Note inlet restriction:

Task completed ___

4. List moisture drain location(s) and check for plugged drain.

Task completed ___

5. Student to observe and document visual air/dirt seepage locations. List all.
   - Air filter canister seal ___
   - Intake pipe clamps (check torque as per OEM specs) ___
   - Check canister for damage ___

Task completed ___

6. Student to perform the charge air cooler test procedure. Plug at inlet and exit, then charge with regulated shop air at OEM spec (this will range from 5 psi to a maximum of 35 psi). DO NOT overpressurize. Check bleed-down time to specification. ___

Task completed ___

Also perform the following:

- Check CAC for flow direction markings. ___
- Remove and inspect turbocharger pressure side pipe and note possible oil residue. ___
- If oil residue is present, remove and inspect turbocharger oil return line for obstruction (coking). ___
- Use a dial indicator to check for turbine shaft endplay ___ and radial play. ___

Task completed ___

Instructor Check/Comments

____

____

____

INTERNET TASKS

Research how different OEMs are using EGR in their post-2010 diesel engines equipped with SCR systems. Log on to the Internet, and use a search engine to access the websites of the following manufacturers. This is one instance where access to an OEM online SIS would be an advantage.

- Caterpillar®
- Cummins®
- Freightliner®, Mercedes-Benz®
- Volvo-Mack®
- Navistar®
- Paccar®
- Hino™, Toyota®

STUDY TIPS

Identify five key points in Chapter 9.

Key point 1 ______________________

Key point 2 ______________________

Key point 3 ______________________

Key point 4 ______________________

Key point 5 ______________________

CHAPTER 10

Engine Brakes

OBJECTIVES

After studying this chapter, you should be able to:

- Identify some of the different types of engine brakes used on highway diesel engines.
- Describe the operating principles of each type of engine brake.
- Outline the controls used to manage engine brakes.
- Describe how the hydraulic actuation of internal engine compression brakes is managed and timed.
- Explain how a progressive, multicylinder engine braking is managed.
- Describe the operation of a constant throttle valve (CTV) brake.
- Outline the differences in automatic and manual control of the Caterpillar BrakeSaver.

END OF CHAPTER REVIEW QUESTIONS

1. Which type of engine brake uses the compression stroke of the piston as its retarding stroke?
   a. Internal compression brake
   b. External compression brake
   c. Hydraulic retarder
   d. Exhaust pressure governor
2. Which type of engine brake uses the exhaust stroke of the piston as its retarding stroke?
   a. Internal compression brake
   b. External compression brake
   c. Hydraulic retarder
3. Technician A says that some newer six-cylinder engines option engine braking on just one engine cylinder. Technician B says that progressive engine braking on newer engines can option engine braking on one to all six cylinders. Who is correct?
   a. Technician A only
   b. Technician B only
   c. Both A and B
   d. Neither A nor B
4. An engine brake on most current diesel engines is switched by which of the following?
   a. Electromechanically by the driver using a dash switch
   b. Electromechanically by the driver using a pedal switch
   c. By the ECM based on input signals
   d. By a dedicated electronic controller networked to the data bus

5. When will any kind of engine brake usually operate at peak efficiency?
   a. At idle rpm
   b. At peak torque rpm
   c. At rated speed rpm
   d. At the highest rpm
6. Technician A states that on most internal engine compression brakes, the effective braking stroke of the piston is the compression stroke. Technician B states that on some engine brake systems, both the compression and exhaust strokes can deliver engine braking. Who is correct?
   a. Technician A only
   b. Technician B only
   c. Both A and B
   d. Neither A nor B
7. The operating principle of an internal engine compression brake is to convert the engine into an
   a. energy-absorbing compressor.
   b. energy-releasing machine.
   c. inertia-absorbing device.
   d. inertia-producing device.
8. When an engine brake is used, what is the energy of vehicle motion ultimately converted to?
   a. Kinetic energy
   b. Heat energy
   c. Chemical energy
   d. Potential energy
9. When the term "Jake brake" is used, which of the following is usually being referred to?
   a. Internal compression brake
   b. Constant throttle valve
   c. External compression brake
   d. Driveline retarder
10. When a Caterpillar BrakeSaver is used, which of the following is also true?
    a. External exhaust brake principles apply
    b. A larger capacity engine oil sump is used
    c. It must be controlled by the vehicle ECM
    d. Braking efficiency can never be modulated

INTERNET TASKS

Log on to the Internet and use a search engine to research the following terms and record any information you can find about diesel engine brakes.

- Jacobs Vehicle Systems (try "Jake brake" as well)
- Pacbrake
- Williams brake
- Mercedes-Benz engine brakes
- Caterpillar BrakeSaver
- Cummins InteBrake

STUDY TIPS

Identify five key points in Chapter 10.

Key point 1 ______________________________

Key point 2 ______________________________

Key point 3 ______________________________

Key point 4 ______________________________

Key point 5 ______________________________

CHAPTER 11

Servicing and Maintenance

OBJECTIVES

After studying this chapter, you should be able to:

- Explain why it is sometimes important to connect to a chassis data bus even when performing routine service work.
- Outline the procedure required to break in a diesel engine.
- Respond appropriately when a diesel fuel tank gets filled with gasoline.
- Service the air intake system.
- Identify the appropriate oil to use in a diesel engine.
- Identify the steps required to perform an engine service.
- Perform an engine oil and filter change.
- Mix and test coolant, and service an engine cooling system.
- Identify the appropriate fuel and refuel/prime a diesel engine.
- Respond to a water-in-fuel alert and service a water separator.
- Service and replenish an SCR system with DEF.
- Identify when it is appropriate to service a DPF.

END OF CHAPTER REVIEW QUESTIONS

1. Which of the following engine lubricants should be used if an engine is equipped with a DPF?
    a. Any synthetic lubricant
    b. Any biodiesel lubricant
    c. API CI-4
    d. API CJ-4
2. Which of the following would best describe the color of DEF?
    a. Blue
    b. Clear
    c. Amber
    d. Red

3. What quantity of oil does the distance from the top to the bottom of the crosshatch area on the dipstick of a typical heavy-duty, commercial diesel engine represent?
   a. 1.0 quart (0.95 L)
   b. 2.0 quarts (1.9 L)
   c. 4.0 quarts (3.8 L)
   d. 8.0 quarts (7.5 L)
4. Technician A says that the fastest way to break in a light-duty diesel engine is to add a charge of aftermarket break-in solution to the engine oil. Technician B says that manufacturers recommend adding fuel conditioners to diesel fuel tanks to improve the winter driveability of diesel engines. Who is correct?
   a. Technician A only
   b. Technician B only
   c. Both A and B
   d. Neither A nor B
5. A diesel fuel tank has been filled with gasoline and run until it quits. Which of the following should be done?
   a. Flush fuel system with diesel fuel
   b. Replace the fuel filter(s)
   c. Remove and service the fuel injectors
   d. All of the above
6. Which of the following solutions would provide the highest level of antifreeze protection?
   a. Pure distilled water
   b. A 50:50 solution of EG and distilled water
   c. A 100 percent solution of PG
   d. A 100 percent solution of EG
7. What quantity of oil does the distance from the top to the bottom of the crosshatch area on a typical light-duty, automotive diesel engine represent?
   a. 1.0 quart (0.95 L)
   b. 2.0 quarts (1.90 L)
   c. 4.0 quarts (3.80 L)
   d. 8.0 quarts (7.60 L)
8. Technician A says that DEF is available at most diesel refueling stations. Technician B says that DEF can be obtained from most manufacturer dealerships. Who is correct?
   a. Technician A only
   b. Technician B only
   c. Both A and B
   d. Neither A nor B
9. Which of the following is the correct course of action when a WIF alert is posted to the DDU?
   a. Drain the water separator sump
   b. Replace the fuel filters
   c. Replace the water separator
   d. All of the above
10. When should a diesel engine air filter be changed?
   a. At every engine wet service
   b. Every time the oil filter is changed
   c. Every six months
   d. When the restriction gauge indicates it is beginning to plug

JOB SHEET 11.1

Name ______________________________ Date ______________

Job Description: Change engine oil and filters.

Performance Objective: Remove and replace the engine oil and filter using the manufacturer's guidelines and check the air filter performance.

Text Reference: Chapter 11.

Protective Clothing: Required for shop exercise sessions: hearing, eye, hand, and foot protection, and shop coat/coveralls.

Tools and Materials:

Diesel engine
Wrenches
Catch-pan
New filters
O-ring for sump plug
Appropriate quantity of new engine oil

PROCEDURE

1. Log onto the chassis data bus where required to record the service event.

Task completed ______________

2. Check air inlet restriction using the filter minder scale or equivalent. Note the specifications. Filters are usually changed on a calendar (time) basis, not at every service, so if the inlet restriction specification is OK, leave it. Change the air filter if the reading exceeds specifications.

Task completed ______________

3. Check engine oil level and make note of it. If oil level is high, it could be contaminated by another fluid (coolant = milky appearance, fuel = black color), in which case further repairs are required. If oil level is low, note on the work order because this may indicate engine damage. If the oil level reads within the normal range on the dipstick, proceed to drain the oil. **Caution:** Oil may be hot. If an oil sample is to be taken for analysis, do so while the oil is in mid-flow from the oil pan during draining. See **Figure 11–1** and **Figure 11–2**.

Task completed ______________

FIGURE 11–1 Loosening an oil filter(s) with an oil filter wrench.

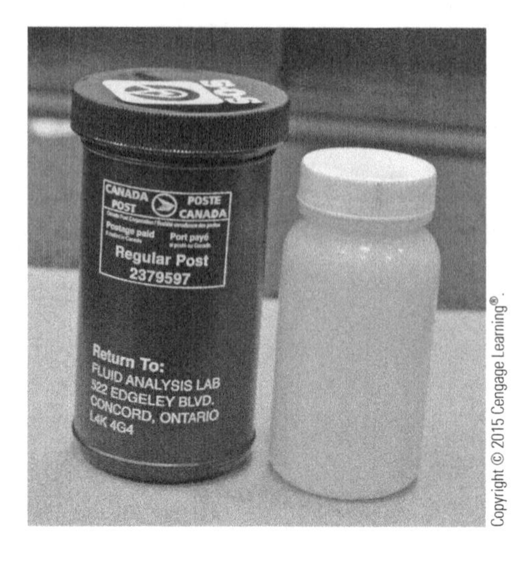

FIGURE 11–2 Mail-in oil sample bottle favored by Caterpillar dealerships.

4. Remove and replace the oil filter(s). Most HD commercial diesel engines use more than one oil filter, and although they may be similar in appearance, a combination of full flow and centrifugal filters may be used. Lightly smear clean oil over the filter sealing gasket and torque exactly according to the instructions.

   Task completed ________________

5. Replace seating washer or O-ring on oil pan drain plug. Reinstall drain plug and torque to specifications.

   Task completed ________________

6. Refill with new engine oil to the appropriate specifications.

   Task completed ________________

7. Reset the smart oil life service indicator if equipped (may require OEM electronic service tool EST).

   Task completed ________________

8. Start engine; idle for two minutes while checking for leaks. Run to high idle for 15 seconds. Shut down the engine and after waiting a minimum of five minutes, check and correct the oil level.

   Task completed ________________

Instructor Check/Comments

____

____

____

JOB SHEET 11.2

Name ____ Date ____________________

Job Description: Coolant test and coolant filter replacement.

Performance Objective: Use the OEM procedure to check the coolant chemistry and replace the coolant filter.

Text Reference: Chapter 11.

Protective Clothing: Required for shop exercise sessions: hearing, eye, hand, and foot protection, and shop coat/coveralls.

Tools and Materials:

Refractometer
Hydrometer
New coolant filter

PROCEDURE

1. Check coolant chemistry using the appropriate OEM test strips and measure the antifreeze protection with a refractometer (preferred) or hydrometer. See **Figure 11–3.**

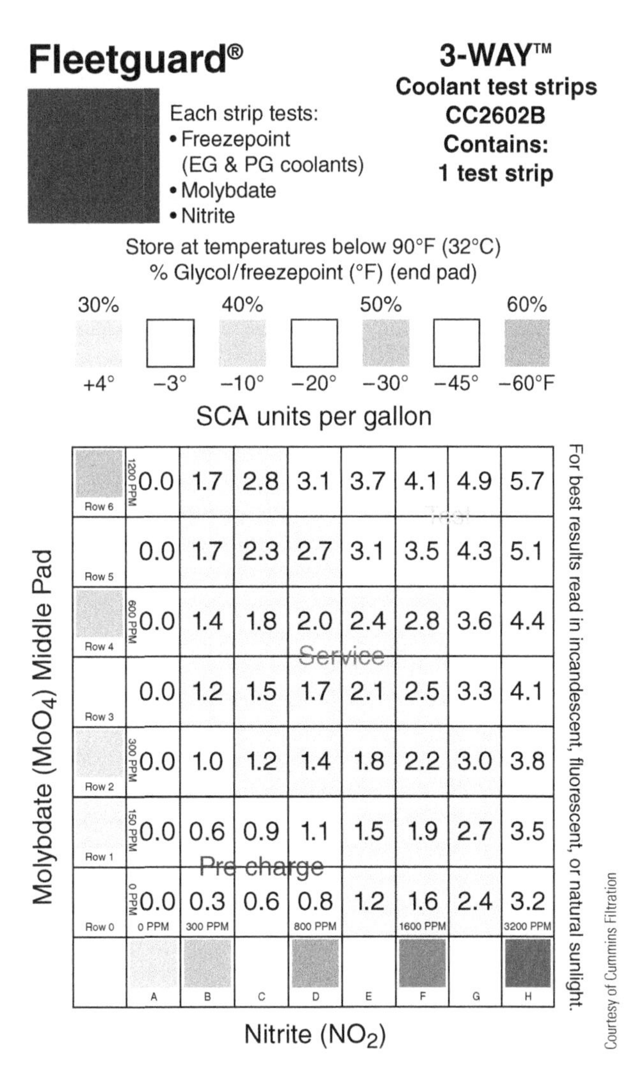

FIGURE 11–3 Test card used with a three-way Fleetguard coolant test strip.

Task completed ______________

2. Does the antifreeze concentration need to be corrected, yes or no?

Task completed ________________

3. Replace the coolant filter. (Sometimes OEMs require coolant filters to be replaced over logged engine hours or on a calendar schedule because the additive chemical pack is loaded into the filter canister.)

Task completed ________________

4. Check the radiator and surge tank coolant levels.

Task completed ________________

5. Check the radiator fins for dirt/clogging.

Task completed ________________

6. Check the fan blades for nicks/damage and the fan shrouding for damage.

Task completed ________________

7. At the completion of the cooling system service, start the engine and check for leaks.

Task completed ________________

Instructor Check/Comments

____

____

____

JOB SHEET 11.3

Name ______________________________ Date ______________

Job Description: Fuel filter servicing.

Performance Objective: Replace the fuel filters on a diesel engine and prime the fuel system to start engine and check for leaks.

Text Reference: Chapter 11.

Protective Clothing: Required for shop exercise sessions: hearing, eye, hand, and foot protection, and shop coat/coveralls.

Tools and Materials:

Diesel engine
Set of fuel filters
Wrenches

PROCEDURE

Each diesel fuel system is distinctive, so it is recommended that the OEM service literature be consulted. The procedure outlined in this job sheet is general and it will not necessarily work on all vehicles.

1. Check and service the water separator. Drain any water present. Reset the WIF if necessary. See **Figure 11–4.**

FIGURE 11–4 A Fleetguard Diesel Pro multifunction fuel filter/water separator assembly.

Task completed ______________

2. Remove and drain the primary filter, followed by the secondary filter, observing environmental regulations. Agricultural equipment that has gravity feed to the primary filter may have a drain valve that needs to be shut off before beginning this procedure.

Task completed ______________

3. Replace the primary filter and prime the filter. Use the approved method of priming the filter, which usually requires pouring clean, filtered fuel into the inlet. When filled, fit the filter to its mounting pad and torque to specification.

Task completed ______________

4. Install the secondary filter dry and then prime. Today most OEMs recommend that the secondary filter be installed dry and primed on the engine using one of two methods. (1) *Hand primer pump:* Usually located on or close to the transfer pump. Loosen the coupling to the exit line on the secondary pad. Actuate the primer pump until bubble-free fuel exits the coupling. When this happens, snug off the coupling. (2) *Electric primer pump:* Some OEMs use automatic priming by electric pump. Use the OEM instructions to actuate the priming circuit. **Caution:** Today's high-pressure common rail (CR) fuel systems are highly sensitive to ANY contamination, and failure to observe the OEM priming procedure can result in costly injector damage.

Task completed ______________

5. After priming, crank and start the engine. When rpm roll ceases, run the engine rpm briefly up to high idle rpm and back to idle. Check for leaks.

Task completed ______________

Instructor Check/Comments

____

____

____

JOB SHEET 11.4

Name ______________________________ Date ______________

Job Description: Check diesel external emission devices.

Performance Objective: Use an EST to check the operation of the SCR circuit and check/replenish DEF.

Text Reference: Chapter 11.

Protective Clothing: Required for shop exercise sessions: hearing, eye, hand, and foot protection, and shop coat/coveralls.

Tools and Materials:
Diesel engine equipped with an SCR circuit
EST capable of reading fault codes
DEF

PROCEDURE

1. Check engine ECM active and historical codes and note whether any SCR-related codes have been logged. Determine whether corrective action is required by consulting the OEM service literature.

Task completed ______________

2. Check and replenish DEF level. When refilling a DEF reservoir that has been emptied enough to set a code, there may be a delay before the ECM registers the new level. Handle DEF with some care, noting that it is corrosive to paint and should never be put into a fuel tank. Be mindful of the weather conditions and note that in winter temperatures DEF can freeze. See **Figure 11–5**.

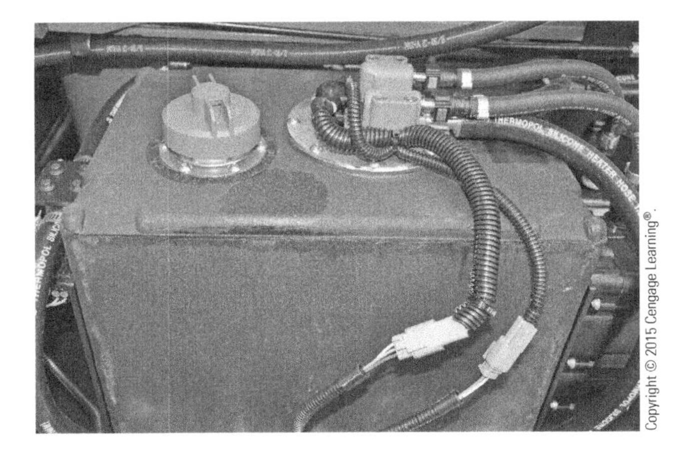

FIGURE 11–5 A typical DEF tank used on a post-2010 highway truck.

Task completed ______________

3. Start the engine. Ensure that no DEF-related codes are active, and if any historical codes were set, erase them.

Task completed ______________

Instructor Check/Comments

____

____

____

JOB SHEET 11.5

Name ______________________________ Date ______________

Job Description: DPF service.

Performance Objective: Check DPF operation and determine whether a regeneration event is required.

Text Reference: Chapter 11.

Protective Clothing: Required for shop exercise sessions: hearing, eye, hand, and foot protection, and shop coat/coveralls.

Tools and Materials:

Diesel engine equipped with a DPF circuit
EST capable of reading fault codes
OEM service literature
Standard shop tools

PROCEDURE

1. With engine key-on but not running, log onto the chassis data bus and use an EST to check for any DPF-related fault codes. Note when the previous active regeneration took place.

   Task completed ______________

2. If the code profile indicates that an active DPF regeneration is required, prepare the engine/truck. In most cases, this will require removing the unit from an open service facility environment. **Caution:** Never attempt to regenerate a DPF by connecting to a standard shop air extraction system.

   Task completed ______________

3. After the regeneration, cool engine, then recheck for any emissions-related fault codes. Clear historical codes.

   Task completed ______________

Instructor Check/Comments

____

____

____

INTERNET TASKS

1. Check out the Cummins, Detroit Diesel, John Deere, Volvo, and Caterpillar websites and locate whatever information you can on each company's service information systems.
2. Search the following key words: diesel exhaust fluid (DEF), B5 biodiesel, B20 biodiesel. See if you can locate the MSDS for DEF.
3. Check out the Volkswagen, Audi, BMW, and Nissan websites and make a list of each company's preferred diagnostic ESTs, software, and service information systems. See if you can note any differences from the commercial engine websites listed in task 1.
4. Investigate the testimonial evidence (online chat) of the consequences of using a non–CJ-4 engine oil in an engine equipped with a DPF.
5. Investigate the testimonial evidence (online chat) of the consequences of putting substances other than DEF into an SCR tank.

STUDY TIPS

Identify five key points in Chapter 11.

Key point 1 ______________________________

Key point 2 ______________________________

Key point 3 ______________________________

Key point 4 ______________________________

Key point 5 ______________________________

CHAPTER 12

Engine Removal, Disassembly, Cleaning, Inspection, and Reassembly Guidelines

OBJECTIVES

After studying this chapter, you should be able to:

- **Remove an engine from a typical truck chassis.**
- **Disassemble a typical diesel engine.**
- **Outline the process of cleaning and inspecting engine components.**
- **Tag and organize components and connectors during engine disassembly.**
- **Describe some key reconditioning procedures.**
- **Develop good inspection and failure analysis habits.**
- **Evaluate components for repair or replacement.**
- **Describe the procedure required to reassemble a diesel engine.**

END OF CHAPTER REVIEW QUESTIONS

1. When hoisting a crankshaft from an inverted six-cylinder engine block using a yoke, which of the following would be the preferred location to attach the yoke hooks?
   a. Throw journals numbers 2 and 5
   b. Throw journals numbers 3 and 4
   c. Main journals numbers 1 and 6
   d. Main journals numbers 3 and 4

2. When measuring piston ring end gap, which of the following should be true?
   a. The piston should be at operating temperature
   b. The ring and liner should be at operating temperature
   c. The ring and liner should be at room temperature
   d. The ring should be installed into its piston ring groove
3. While indicating flywheel housing concentricity on a housing specified to be within 0.008 inch, you record the following measurements:

NE	SE	SW	NW	NE
0.000 in.	+0.003 in.	+0.005 in.	–0.004 in.	–0.004 in.

   What should you do?
   a. Calculate that the TIR is within specification.
   b. Repeat the procedure, because the indicator has moved.
   c. Calculate that the TIR is close enough to spec to leave as is.
   d. Remove dowels and reset flywheel housing concentricity.
4. While indicating flywheel housing concentricity on a housing that must be within 0.012 inch TIR, you record the following measurements:

NE	SE	SW	NW	NE
0.000 in.	+0.006 in.	+0.003 in.	–0.005 in.	–0.000 in.

   What should you do?
   a. Calculate that the TIR is within specification.
   b. Repeat the procedure; the indicator has moved.
   c. Calculate that the TIR is out of specification.
   d. Remove dowels and reset flywheel housing concentricity.
5. When calculating true TDC on a diesel engine, which of the following instruments should be used?
   a. Straightedge
   b. Micrometer
   c. Feeler gauges
   d. Dial indicator
6. When measuring cylinder block deck warpage, which of the following tools should be used?
   a. Straightedge and feeler gauges
   b. Micrometer
   c. Depth gauge
   d. Dial indicator
7. Which of the following tools would help you to measure shaft runout on an out-of-engine camshaft?
   a. V-blocks
   b. Micrometer
   c. Feeler gauges
   d. Depth gauge
8. What is the purpose of a preluber?
   a. Increase lube circuit flow
   b. Lube the oil circuit before first startup
   c. Clean used engine oil
   d. Recycle engine coolant

9. How should a crankshaft be stored after removing it from a cylinder block if it is to be reused?
   a. Horizontally on the floor
   b. Should never be reused
   c. Standing upright on the floor
   d. Safely secured in a vertical position
10. Technician A says that turbocharger gaskets should never be reused after rebuilding an engine. Technician B says that turbocharger studs should always be replaced when an engine is reconditioned. Who is correct?
    a. Technician A only
    b. Technician B only
    c. Both A and B
    d. Neither A nor B

JOB SHEET 12.1

Name ______________________ Date ______________

Job Description: Indicate flywheel housing concentricity.

Performance Objective: After completing this assignment, the student should be able to check the radial concentricity of a flywheel housing to the crankshaft using the OEM procedure and specifications.

Text Reference: Chapter 12.

Protective Clothing: Required for shop exercise sessions: hearing, eye, hand, and foot protection, and shop coat/coveralls.

Tools and Materials:

Assembled diesel engine on a stand with OEM service specifications
Engine barring tool
Magnetic base dial indicator capable of reading 0.001 inch
Chalk
5-pound rubber mallet
Plug reamers
Basic hand tools

PROCEDURE

You should refer to **Figure 12–1** to complete this exercise. This Caterpillar method uses reference points A, B, C, D rather than the NE, NW, SW, SE we have recommended in the core text. The student is to:

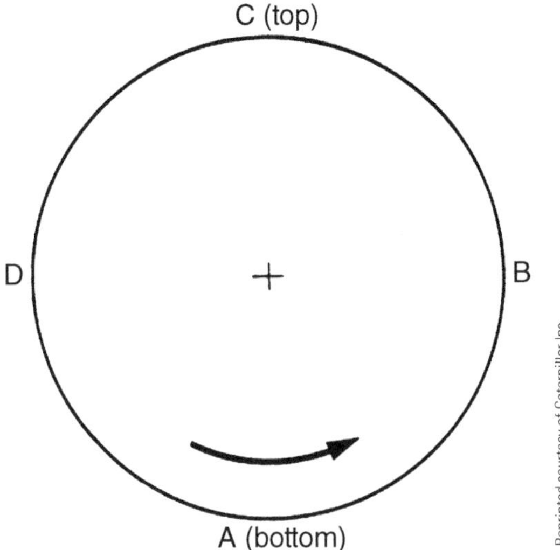

FIGURE 12–1 Checking flywheel housing radial concentricity.

1. Check OEM for the radial flywheel housing maximum TIR specification.

   Task completed ________________

2. Remove the flywheel from the crankshaft. Clean the inner flange bore of the flywheel housing using fine-grade emery cloth.

   Task completed ________________

3. Mark the flywheel face of the engine with the chalk as specified in the OEM literature. If the location is not specified, mark at four points as follows: NE, NW, SW, SE.

   Task completed ________________

4. Mount the dial indicator securely at any point on the crankshaft. Position the probe on the inner flange bore at one of the four chalk-marked location indices, e.g., NE. Make sure that the indicator plunger is loaded through at least one full turn of the indicator dial, then set at zero. Gently, manually bar the engine through one complete rotation; the indicator dial should read exactly zero after the full turn. If it does not, check the indicator base, which has probably moved on the crankshaft.

   Task completed ________________

5. Bar the engine through a second rotation, this time recording the runout value either above or below the zero point on the dial indicator. The highest positive value should be added to the lowest negative value to calculate the TIR. If the following readings were obtained:

NE	0
NW	+0.003
SW	+0.005
SE	–0.002

   the 0.005 inch reading is added to the –0.002 inch reading to produce a TIR value of 0.007 inch. (If you find the result of this calculation to be confusing, please refer to Chapter 12 of the core text for a full explanation of how TIR is calculated.)

   Task completed ________________

   To correct:

   1. Remove the flywheel housing. Drive out the interference fit alignment dowels; there are usually two, offset with one on each side of the flywheel housing.
   2. Reinstall the flywheel housing, but leave the fasteners snug to the extent that the flywheel housing can be moved on the cylinder block when struck with a 5-pound rubber mallet.
   3. Perform the checking procedure again. Use the rubber mallet to reposition the flywheel housing until the OEM-specified TIR is met. Then torque the flywheel fasteners to specification.
   4. Measure the alignment dowels. Select the next size up oversize alignment dowels. Use a plug reamer to ream the engine cylinder to the required dimension of the oversize alignment dowel minus the recommended interference fit.
   5. Thoroughly clean the dowel bores and very lightly lubricate the dowels. Use a 3-pound hammer and drive the dowels flush.

   Task completed ________________

Instructor Check/Comments

____

____

____

JOB SHEET 12.2

Name ______________________________ Date ______________

Job Description: Perform a diesel engine compression test.

Performance Objective: After completing this assignment, the student should be able to perform a diesel engine cylinder compression test and compare the results to the OEM service specifications.

Text Reference: Chapter 12.

Protective Clothing: Required for shop exercise sessions: hearing, eye, hand, and foot protection, and shop coat/coveralls.

Tools and Materials:
Injector removal tooling
Compression testing gauge and injector bore adaptors
OEM service specifications

PROCEDURE

Testing engine cylinder pressures is less commonly practiced today than it might have been 20 years ago because there are usually faster, more accurate means of diagnosing cylinder malfunctions. However, diesel technicians should understand the procedure required to test cylinder pressure. The procedure outlined below attempts to describe the methods used to test hydraulic and unit injector-fueled engines.

1. Disable fuel system. The means used to accomplish this will vary with the fuel system. If the injectors are supplied with fuel from a fuel rail in the cylinder head, it is *essential* that the supply and return galleries are drained; if not, this fuel will drain on top of the pistons when the injectors are pulled.

   Task completed ______________

2. Pull all of the injectors. Cylinder pressure data will vary according to engine rpm (the faster the rpm, the less time for leakage), and test specifications usually require that the only cylinder under pressure is the one being tested. Make sure that the batteries are fully charged.

   Task completed ______________

3. Fit the adaptor to the #1 engine cylinder and the gauge to the adaptor plumbing. Crank the engine through several revolutions (three or four) and record the peak pressure value for that cylinder. Repeat the process, working through the engine cylinders.

Engine Cylinder	Peak Cylinder Pressure	Cranking Speed
#1		
#2		
#3		
#4		
#5		
#6		
#7		
#8		

   Task completed ______________

4. In analyzing the results of a cylinder pressure test, the data must meet two criteria: the specification window and balance window. That is, all the pressure values recorded must fall within the OEM specification range, and, additionally, must be within a certain amount of each other.

Task completed ______

Instructor Check/Comments

______

______

______

INTERNET TASKS

Log on to the Internet and use a search engine to research the following topics. See if you can identify any environmental concerns over engine cleaning processes.

1. MagnaFlux®
2. Magnetic-flux testing
3. Hot soak tanks
4. Alkaline solutions for boiling cylinder blocks
5. Water separator sluices

STUDY TIPS

Identify five key points in Chapter 12.

Key point 1 ______

Key point 2 ______

Key point 3 ______

Key point 4 ______

Key point 5 ______

CHAPTER 13

Fuel Subsystems

OBJECTIVES

After studying this chapter, you should be able to:

- Identify fuel subsystem components on a typical diesel engine.
- Describe the construction of a fuel tank.
- Explain the operation of and troubleshoot a fuel sending unit.
- Define the role of primary and secondary fuel filters.
- Service primary and secondary fuel filters.
- Explain how a water separator functions.
- Service a water separator.
- Define the operating principles of a transfer pump.
- Prime a fuel subsystem.
- Test the low-pressure side of the fuel subsystem for inlet restriction.
- Test the charge side of the fuel subsystem for charging pressure.
- Identify some typical sensors used in diesel fuel subsystems.

END OF CHAPTER REVIEW QUESTIONS

1. On the typical truck diesel fuel subsystem, which of the following is at the lowest pressure in the circuit?
   a. Fuel heater
   b. Primary filter
   c. Secondary filter
   d. Charging circuit
2. Where is a secondary filter located?
   a. Upstream from the transfer pump
   b. On the charge side of the transfer pump
   c. In the fuel rail
   d. In the return gallery
3. What is the main reason for filling vehicle fuel tanks before overnight parking?
   a. To minimize moisture condensation in the tanks
   b. To minimize fuel evaporation
   c. To help cool down on-board fuel
   d. Drivers may forget the next morning

4. Besides fuel storage, the fuel tank may play an important role in a high-flow fuel system as a(n)
   a. heat exchanger.
   b. fuel heating device.
   c. ballast equalizer.
   d. aerodynamic aid.
5. Which of the following could be used to test the low-pressure side of fuel subsystems for inlet restriction?
   a. Diagnostic sight glass
   b. H_2O manometer
   c. Negative pressure gauge
   d. Accurate high-pressure gauge
6. Which of the following should fuel subsystem charging pressures be measured with?
   a. Diagnostic sight glass
   b. H_2O manometer
   c. Hg manometer
   d. Accurate pressure gauge
7. What should be used to check for air being pulled into a fuel subsystem?
   a. Diagnostic sight glass
   b. H_2O manometer
   c. Hg manometer
   d. Accurate pressure gauge
8. Which type of fuel transfer pump is most commonly used by electronically managed engine/fuel systems?
   a. Plunger pump
   b. Centrifugal pump
   c. Diaphragm pump
   d. Gear pump
9. Which type of pump is used as a typical hand primer pump?
   a. Single-acting plunger
   b. Double-acting plunger
   c. Rotary gear
   d. Cam-actuated diaphragm
10. Charging pressure rise is usually directly related to which of the following?
    a. Throttle position
    b. Engine load
    c. Peak power
    d. Increased rpm

JOB SHEET 13.1

Name ______________________________ Date ______________

Job Description: Prime a fuel subsystem.

Performance Objective: After completing this assignment, the student should be able to follow the correct procedure to start up an engine that has run out of fuel.

Text Reference: Chapter 13.

Protective Clothing: Required for shop exercise sessions: hearing, eye, hand, and foot protection, and shop coat/coveralls.

Tools and Materials:

Hand primer pump
Supply of clean fuel
OEM service literature

PROCEDURE

The following procedure is used to prime a typical fuel subsystem; obviously, each fuel system is distinct, so the OEM service literature should be consulted. You can refer to **Figure 13–1** to locate the components of a fuel subsystem. If the reason for priming the fuel subsystem is a shutdown caused by water plugging the secondary fuel filter, then *both* the primary and secondary fuel filters should be replaced.

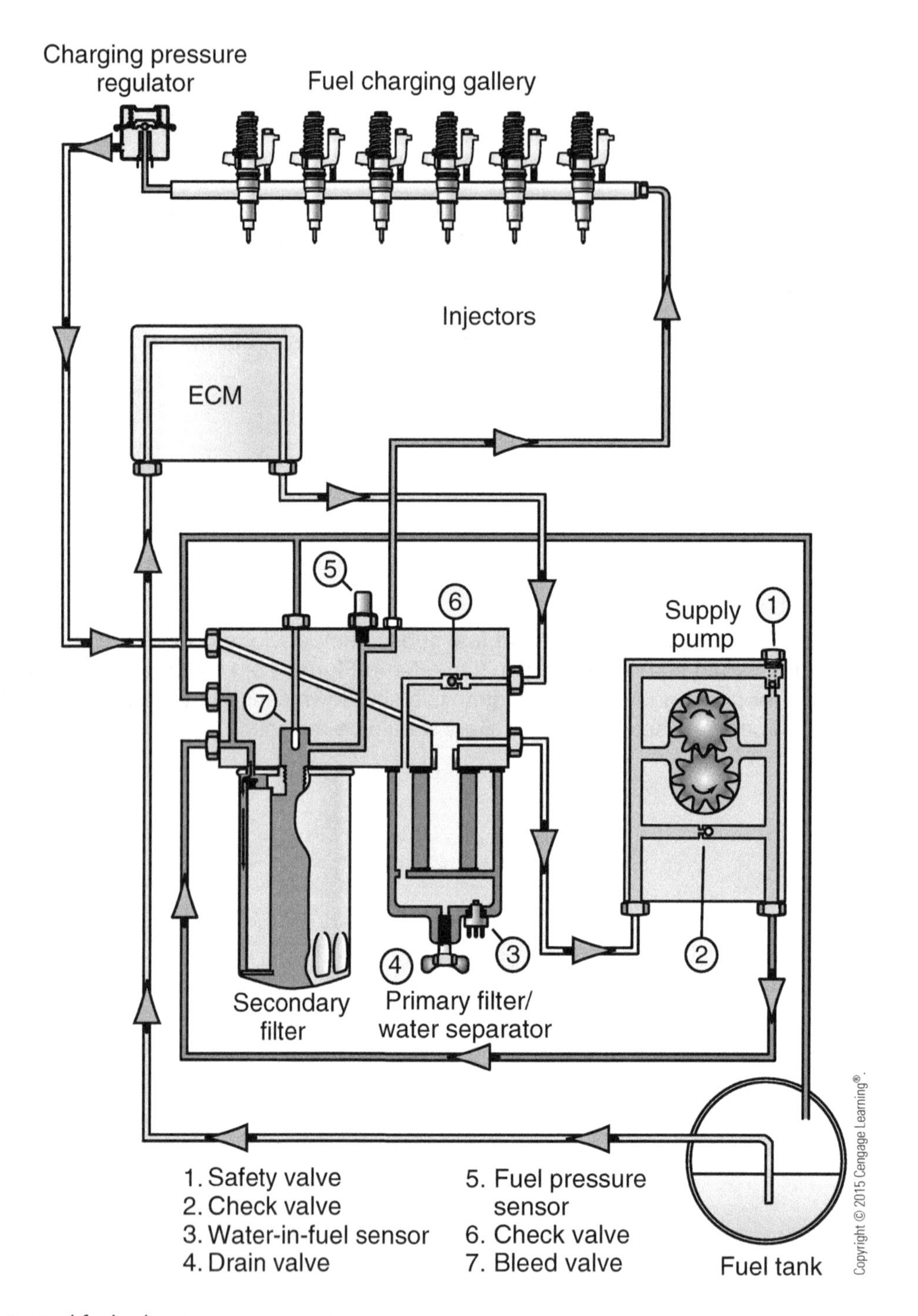

FIGURE 13–1 Typical fuel subsystem components.

1. Make sure that there is fuel in the vehicle fuel tanks.

   Task completed ________________

2. Remove the primary filter. Carefully pour filtered fuel into the inlet (outer) annulus until the fuel level in the outlet aperture is close to full and is level with the inlet level. Install the primary fuel filter.

   Task completed ________________

3. Fit a hand primer pump on the suction side of the fuel transfer pump.

   Task completed ________________

4. In some systems, it may be possible to actuate the hand primer pump and charge the secondary filter with fuel; this is the preferred method because it reduces the chance of contaminating the fuel. If the secondary fuel filter has to be removed, make sure that no contaminants enter the fuel filter. Always pour the fuel into the inlet annulus of the filter.

   Task completed ________________

5. Locate a bleed point in the system. In an inline injection pump this would be the exit port from the charging gallery. In engines that use fuel supply and return manifolds in the cylinder head, the appropriate bleed point may be at the inlet or outlet of either line. Crack the line nut.

   Task completed ________________

6. Actuate the hand primer pump. This should result in fuel exiting the cracked line nut and fitting. Initially the fuel will be aerated. Pump until the fuel appears bubble-free, then retorque the line nut.

   Task completed ________________

7. Crank engine with accelerator pedal fully depressed in 30-second phases until the engine starts. Observe the stack(s) for smoke during cranking. Remember, smoke indicates that the fuel subsystem is probably charged.

   Task completed ________________

Instructor Check/Comments

____

____

____

INTERNET TASKS

Search the Internet for information on aftermarket suppliers of fuel filters, heaters, and water separators, and compare their information with that provided by the OEM. Enter the following manufacturers into your search engine to get started; you should be able to add to this list:

1. Fleetguard®
2. Davco®
3. Donaldson®
4. Baldwin®
5. Purolator®

STUDY TIPS

Identify five key points in Chapter 13.

Key point 1 ______________________________

Key point 2 ______________________________

Key point 3 ______________________________

Key point 4 ______________________________

Key point 5 ______________________________

CHAPTER 14

Injector Nozzles

OBJECTIVES

After studying this chapter, you should be able to:

- **Identify the subcomponents of a nozzle assembly.**
- **Describe the injector nozzle's role in system pressure management.**
- **Identify two types of injector nozzles.**
- **Describe the principles of operation of multiple-orifice and electrohydraulic nozzles.**
- **Define nozzle differential ratio.**
- **Describe a valve closes orifice (VCO) nozzle.**
- **Bench (pop) test a hydraulic injector nozzle.**
- **Test a nozzle for forward leakage.**
- **Test a nozzle for back leakage.**
- **Outline the procedure required to test an electrohydraulic injector.**
- **Outline the procedure required to remove, inspect, and reconnect high-pressure lines.**

END OF CHAPTER REVIEW QUESTIONS

1. Which type of injector nozzle provides *soft* opening and closing pressures?
   a. Hydraulic multiple-orifice injector
   b. Electrohydraulic injector (EHI)
   c. Any hydraulically switched injector
   d. Any electronically switched injector
2. When injector back leakage is bench checked and is higher than specified, which of the following is the usual cause?
   a. Low NOP setting
   b. High NOP setting
   c. Sticking nozzle valve
   d. Wear
3. When nozzle forward leakage is bench checked, which of the following is being tested?
   a. NOP
   b. The seal at the nozzle seat
   c. Nozzle valve-to-body clearance
   d. Injector spring

4. Replacing a single high-pressure injection line on a multicylinder engine with one of shorter length would likely have what effect on injection timing in the affected cylinder?
   a. Advance
   b. Retard
   c. None
   d. Decrease the fuel pulse width
5. As injection pressure increases to a hydraulic injector nozzle, which of the following is likely to occur?
   a. Injected droplets decrease in size
   b. Injected droplets increase in size
   c. Ignition delay increases
   d. Nozzle valve is more likely to unintentionally close
6. Which type of injector nozzle would be used in most diesel engine fuel systems built after 2007?
   a. Pintle valve
   b. Hard opening value
   c. Multiple-orifice
   d. Electrohydraulic
7. What should be used to remove the nut on a high-pressure injection line?
   a. Torque wrench
   b. Hex socket
   c. Line wrench
   d. Open-end wrench
8. Which of the following is another way of stating the NOP specification?
   a. Not operating properly
   b. Residual line pressure
   c. Peak pressure
   d. Popping pressure
9. Most diesel engine OEMs recommend that injectors be installed into the cylinder head injector bore
   a. dry.
   b. coated with engine oil.
   c. coated with Never-Seeze®.
   d. coated with Lubriplate.
10. Which of the following types of injectors would be best suited to multipulse injection?
   a. Hydraulic poppet
   b. Hydraulic multiorifice
   c. EHI with solenoid actuator
   d. EHI with piezoelectric actuator

JOB SHEET 14.1

Name ______________________________ Date ______________

Job Description: Bench-test a hydraulic injector nozzle.

Performance Objective: After completing this assignment, the student should be able to mount an injector in a hand-actuated injector pop tester and assess the serviceability of a hydraulic injector nozzle.

Text Reference: Chapter 14.

Protective Clothing: Required for shop exercise sessions: hearing, eye, hand, and foot protection, and shop coat/coveralls.

Tools and Materials:

Bench-mounted nozzle pop tester
Diesel test oil
Set of injectors
Spray dispersal template
OEM specifications

PROCEDURE

WARNING: MAKE SURE YOU UNDERSTAND THE DANGERS OF HYDRAULIC PINHOLE INJECTION INJURIES BEFORE UNDERTAKING THIS TASK.

Be sure to wear eye protection during every phase of this workshop procedure. Consult Chapter 14 of the textbook for a more detailed description of the steps within this procedure; observe the OEM specifications and remember that each manufacturer will recommend variations on the sequence outlined here. The idea of performing this test is to follow the sequence until the injector fails one of the tests. At that point, it can be determined that the injector will require further reconditioning. The reconditioning procedure is covered in Chapter 14 of the textbook.

1. Clean exterior of the injector assembly with a brass wire brush. It is generally preferred that the injector not be cleaned on a wire wheel. Mount the injector assembly in the bench pop tester and torque the line nuts. Make sure that the nozzle orifices are directed into the collector canister.

   Task completed ______________

2. Actuate the lever and trigger NOP several times to ensure that there is no air in the nozzle and interconnecting plumbing.

   Task completed ______________

3. Now actuate to NOP three times and record the value. Make sure that the average value and the range of values both meet the OEM specification.

   Task completed ______________

4. Test nozzle forward leakage. This requires loading the test pressure to a small value (say 10 bar) below NOP and holding while observing the nozzle orifices. Trace leakage fails the nozzle.

   Task completed ______________

5. Test nozzle back leakage. This requires loading the test pressure to a small value (10 bar) below NOP and observing the rate of pressure drop-off. Use the OEM specification; a rate of pressure drop-off indicates wear.

Task completed ________________

6. Observe the nozzle spray pattern. This may be performed using a template, but in practice, it is usually performed visually. The dispersal pattern should be balanced.

Task completed ________________

7. A nozzle that has passed the preceding sequence of tests using the manufacturer's test specifications can be returned to service.

Task completed ________________

Instructor Check/Comments

____

____

____

INTERNET TASKS

Search the Internet for information on the environmental and safety concerns when working with high pressure, atomized fuel. Make a list of the specialized personnel and shop equipment required when undertaking injector testing. Enter the following items into your search engine to get started:

1. Siemens AG®
2. Bosch® pump and injector test bench equipment
3. Hartridge pump and injector test bench equipment
4. Caterpillar® injector test equipment
5. Pinhole injection injury

STUDY TIPS

Identify five key points in Chapter 14.

Key point 1 ____________________________

Key point 2 ____________________________

Key point 3 ____________________________

Key point 4 ____________________________

Key point 5 ____________________________

CHAPTER 15

Engine Management Electronics

OBJECTIVES

After studying this chapter, you should be able to:

- Describe the circuit layout of an electronically managed diesel engine.
- Explain what is meant by the input, processing, and output circuits in engine management electronics.
- Identify and explain the operation of some input circuit devices.
- Outline the stages of a computer processing cycle.
- Describe how memory is managed in a diesel engine ECM.
- Identify the different types of memory used in vehicle electronics.
- Define the role played by output circuit devices in a typical diesel engine management system.
- Explain the differences between customer and proprietary data programming.
- Define *multiplexing* and understand how it is used to network the engine electronics with other chassis systems.
- Describe the procedure required to connect an electronic service tool (EST) to the chassis data bus.

END OF CHAPTER REVIEW QUESTIONS

1. Which of the following internal ECM components manages the processing cycle?
   a. CRT
   b. RAM
   c. CPU
   d. ROM
2. Which of the following memory types is volatile?
   a. RAM
   b. ROM
   c. PROM
   d. EEPROM
3. In which memory component would you find the master program for engine management in a typical ECM?
   a. RAM
   b. ROM
   c. PROM
   d. EEPROM

4. To which of the following memory categories would customer data programming be written from an electronic service tool (EST)?
   a. RAM
   b. ROM
   c. PROM
   d. EEPROM
5. Which of the following components conditions V-Ref?
   a. ECM
   b. Voltage regulator
   c. EUI
   d. Personality module
6. What type of data does a thermistor produce?
   a. Pressure data
   b. Temperature data
   c. Rotational speed data
   d. Rotational position data
7. What is a Hall-effect sensor most likely to be used for?
   a. Pressure data
   b. Temperature data
   c. Fluid flow data
   d. Rotational or linear position data
8. What type of information is produced by an induction pulse generator?
   a. Pressure data
   b. Temperature data
   c. Rotational speed data
   d. Altitude data
9. Which of the following components could be used to signal accelerator pedal travel?
   a. Pulse generator
   b. Potentiometer
   c. Thermistor
   d. Variable capacitance sensor
10. Which of the following words best explains what *multiplexing* is on a truck chassis?
    a. Wi-Fi
    b. Handshaking
    c. Networking
    d. Broadband communication

JOB SHEET 15.1

Name ______________________________ Date ______________

Job Description: Perform voltage and resistance measurements on a truck chassis electrical circuit.

Performance Objective: Use a digital multimeter (DMM) to perform voltage and resistance measurements on a truck chassis electrical circuit. Use the DMM readings to calculate and confirm the application of Ohm's Law.

Text Reference: Chapter 15.

Protective Clothing: Required for shop exercise sessions: hearing, eye, hand, and foot protection, and shop coat/coveralls.

Tools and Materials:

A truck with a functional electrical circuit
DMM, shop tools
Calculator

PROCEDURE

This exercise relates to electrical principles in general and is not exclusively confined to engine technology.

1. Record the following vehicle data:

   OEM ____________ Model year ____________ Engine ____________ VIN ____________

   Task completed ____________

2. Select the Volts DC mode on the DMM. If the DMM is not auto-ranging, set it on the 18V scale. Connect the DMM across the battery terminals as shown in **Figure 15–1** and record the voltage value.

   Battery voltage ______________

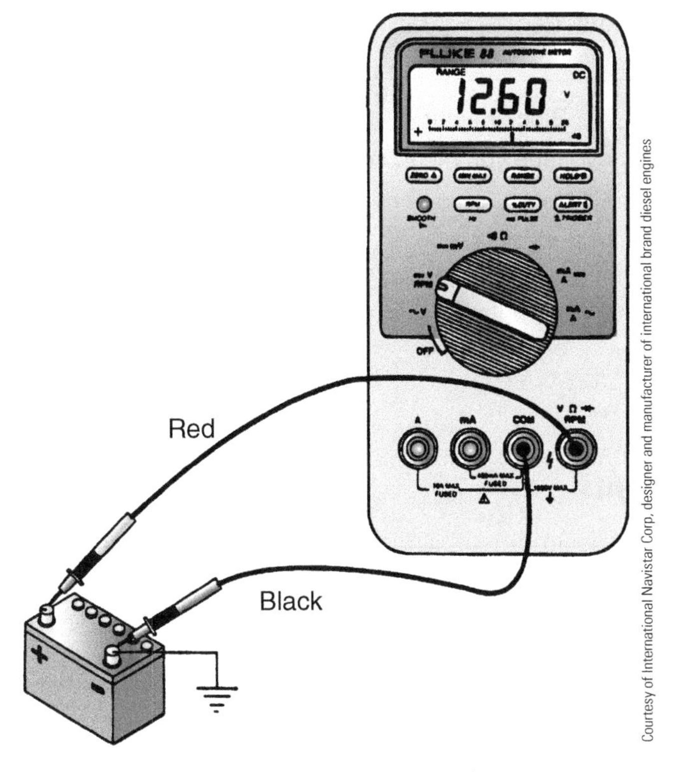

FIGURE 15–1 DMM set up to measure voltage across battery terminals.

Task completed ____________

3. With the engine off, turn on all the dash-controlled electrical loads on the truck and once again record the battery voltage. Turn the electrical loads off.

   Battery voltage ____________________

   Task completed ____________

4. Access one of the truck clearance lights so the terminals on the pigtail can be probed. Do not perform this task if sealed-terminal LEDs are used. Turn the clearance lights on. Ground one DMM probe. Now use the other DMM probe to check the voltage, first on the positive side of the pigtail, then on the negative, recording the values as V1 and V2, respectively.

   V1 ______________ V2 ______________ Sealed LED light circuit ______________

   Task completed ____________

5. Switch off the clearance lights and disconnect the vehicle batteries either by using the isolator switch or by removing the battery terminals.

   Task completed ____________

6. Measure the resistance using the DMM through the clearance light and record the value. Set the DMM into resistance test mode and place one probe on the positive terminal and the other on the negative. Use Ohm's Law and the first battery voltage value you measured (at the beginning of this job sheet) to calculate current flow through the light. Use a calculator. Record the values.

   Clearance light resistance ____________ Battery voltage ____________ Amperage ____________

   Task completed ____________

7. With the batteries still disconnected, measure the resistance values through some typical loads on the truck electrical system. Use those electrical loads that are easiest to access on the truck and take care not to damage any sealed connectors. Record the resistance values.

   Starter motor ____________________ Heater motor ____________________

   Headlight ______________ Block heater ______________ Fuel sending unit ______________

   Using Ohm's Law, the battery voltage you measured in the first test, and a calculator, perform some amperage calculations.

   Task completed ____________

8. Reconnect the truck batteries and verify the operation of any components you have disconnected from the circuit.

   Task completed ____________

Instructor Check/Comments

____

____

____

JOB SHEET 15.2

Name ______________________________ Date ______________

Job Description: Bench-test electronic circuit components.

Performance Objective: Perform resistance tests on components used in truck electronic circuits using a DMM.

Text Reference: Chapter 15.

Protective Clothing: Standard shop apparel, including coveralls or shop coat, safety glasses, and safety footwear.

Tools and Materials:

DMM

A variety of removed electronic components from electronic engine systems, some of which should be functional.

PROCEDURE

Study **Figure 15–2** before beginning this exercise. The objective is to familiarize yourself with using a DMM to perform simple tests that are required on current diesel engines.

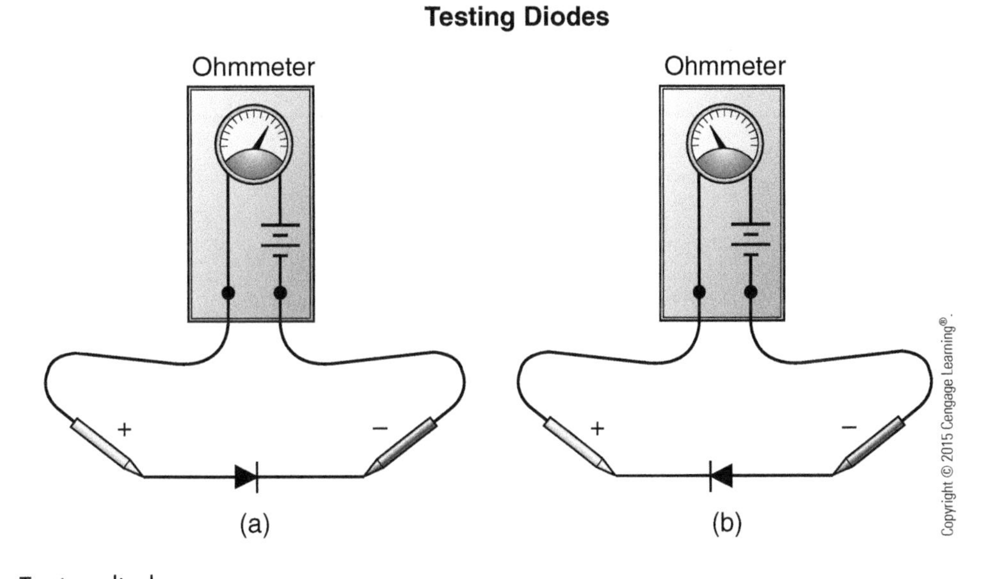

FIGURE 15–2 Testing diodes.

1. Set the DMM in resistance test mode. Where possible, try to obtain the OEM test data for the components to be tested. Many of the electronic circuit components found in training facilities have already failed, so it is important to have some components that are known to function properly.

Task completed ______________

2. Use the following chart to record the OEM-specified values (from service manual/PC download) and the test values you measure. Record the values in Table 15–1.

TABLE 15–1 NTC and PTC Thermistor Performance.

Sensor Component	OEM Resistance	Test Resistance
NTC thermistor at room temperature		
NTC thermistor after immersion in hot water		
Pulse generator sensor		
Piezoresistive pressure sensor		
Variable-capacitance pressure sensor		
Potentiometer		

Task completed ______________

Instructor Check/Comments

____

____

____

JOB SHEET 15.3

Name ______________________________ Date ______________

Job Description: Reprogram a vehicle ECM with proprietary data.

Performance Objective: After completing this assignment, the student should be able to understand the procedure required to reprogram proprietary data files for a vehicle ECM.

Text Reference: Chapter 15.

Protective Clothing: Standard shop apparel, including coveralls or shop coat, safety glasses, and safety footwear.

Tools and Materials:

Truck equipped with an electronically controlled diesel engine
Portable PC equipped with access hardware (communications adaptor [CA]) to the truck data connector
Access to an OEM data hub
OEM reprogramming manual, usually available from the SIS

PROCEDURE

A school or college is only likely to have access to an OEM data hub if it is partnering with OEMs in training programs. If no data hub is available within a training institution, most OEM dealerships will oblige by demonstrating the reprogramming procedure. The procedure outlined is general; each OEM will use variations on it in actual practice. It is good practice to make a copy of customer-programmable data before performing proprietary data reprogramming. The procedure will be slightly different when replacing the engine/vehicle ECM. Remember, any ECM reprogramming requires that the OEM sequencing be adhered to.

1. Download. This step requires a network connection to be made with the OEM data hub using a PC. The PC should be turned on; access the OEM software using the GUI command icons. The truck chassis is usually identified by its VIN. Select the Reprogram ECM option. At this point, the vehicle ECM will normally have to be entered; this is a sort of password for the vehicle. If the objective is to reprogram the engine files (because they have become corrupted or require updating or upgrading), these files are then downloaded into PC memory. This entire reprogramming procedure may be direct if there is a direct handshake connection (hardwire or wireless) between the vehicle electronics and the OEM data hub. However, in most cases it tends to be more convenient to download the engine/vehicle files to PC memory. This is preferable when multiple units require reprogramming. When the file transfer is complete, use the required protocol to back out of and exit the OEM data hub.

   Task completed ______________

2. Reprogram engine/vehicle ECM. In the case of the direct connection between the vehicle ECM and the OEM data hub, steps 2 and 3 in this sequence will occur simultaneously with the first step. However, reprogramming the engine/vehicle files will usually require that the new files downloaded to PC memory be transferred to the ECM via a hardwire or wireless connection with the chassis data bus. Depending on the system, some requirements may have to be met before this file transfer (and overwrite) can be effected. Typical preconditions include the presence of no active codes, key-on, etc. When the file transfer is complete, the data content of the new files in PC memory (now transferred to the ECM) is automatically erased and replaced by a verification file. When wireless reprogramming is performed, the procedure is identical but less involvement by the technician is required; the vehicle driver may not even be aware of what has happened.

   Task completed ______________

3. Upload verification. The verification file is a short file that is automatically created after successful reprogramming of a vehicle ECM. In creating it, the data content of the reprogram files is erased. This file is written to PC memory and should be uploaded to the OEM data hub. Most OEMs require this to be done within a specified time frame after the initial download.

Task completed ________________

4. After performing a proprietary reprogramming sequence, the ignition key usually has to be cycled before startup.

Task completed ________________

Instructor Check/Comments

____

____

____

JOB SHEET 15.4

Name ______________________________ Date ______________

Job Description: Bench-test a transistor.

Performance Objective: After completing this assignment, the student should be able to test semiconductor components in electronic circuits using a DMM. (It should be noted that this is an educational exercise and not one that would be performed on the shop floor.)

Text Reference: Chapter 15.

Protective Clothing: Standard shop apparel, including coveralls or shop coat, safety glasses, and safety footwear.

Tools and Materials:
Selection of transistors
DMM

PROCEDURE

Study **Figure 15–3** before proceeding with the task.

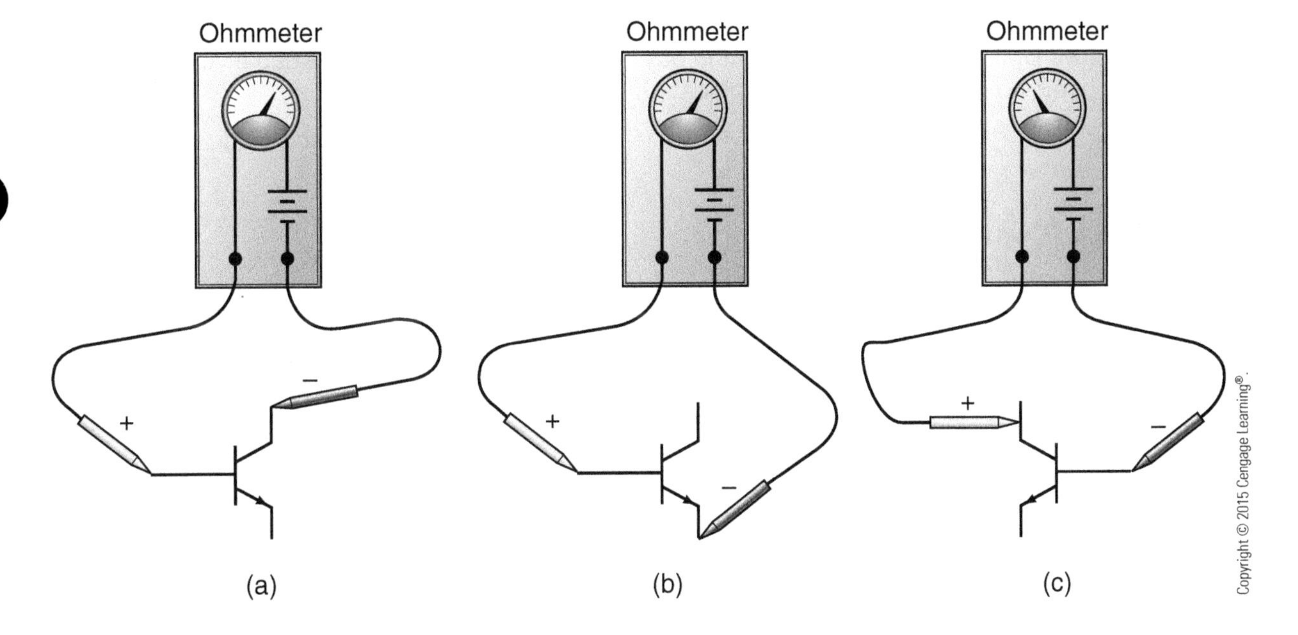

FIGURE 15–3 Testing transistors.

1. Switch on the DMM and select the Resistance test mode. Make sure the polarity of the DMM test leads is known. This can be verified by using a diode: the ohmmeter will indicate continuity through the diode only when the positive lead is connected to the anode and the negative lead is connected to the cathode of the diode.

   Task completed ______________

2. Identify the transistor type. If the transistor is of the NPN type, connect the DMM positive test lead to the base and the negative test lead to the collector. The ohmmeter should read continuity.

   Task completed ______________

3. Now with the positive DMM lead still connected to the base, connect the negative test lead to the emitter. Once again, the ohmmeter should read "continuity," that is, a forward diode junction. If there is no continuity, the transistor is open.

   Task completed ______

4. Connect the negative test lead to the base and the positive test lead to the collector. The reading should indicate no continuity.

   Task completed ______

5. With the negative DMM test lead still connected to the base, reconnect the positive lead to the emitter. There should be no continuity. If high resistance is indicated by the ohmmeter, the transistor may still function in the circuit but is "leaky." If low resistance is indicated, the transistor is shorted.

   Task completed ______

6. To test a PNP transistor, the preceding test sequence is followed but with the DMM test leads reversed in each test. When the negative DMM test lead is connected to the base and the positive test lead is connected to the collector or emitter, a forward diode junction should be indicated.

   Task completed ______

7. When the positive DMM lead is connected to the base of a PNP transistor and the negative test lead to the collector or emitter, no continuity should be indicated.

   Task completed ______

Instructor Check/Comments

______

______

______

JOB SHEET 15.5

Name ______________________________ Date ______________

Job Description: Electronic service tool (EST) orientation.

Performance Objective: After completing this assignment, the student should be able to connect an EST to a vehicle's electronic circuit and perform some of its basic functions.

Text Reference: Chapter 15.

Protective Clothing: Required for shop exercise sessions: hearing, eye, hand, and foot protection, and shop coat/coveralls.

Tools and Materials:

EST. For this exercise either a generic reader/programmer, such as the Nexiq ProLink iQ, or a proprietary reader/programmer can be used, either handheld or stationary.

OEM software specific to the truck or engine electronic system being accessed. If a generic handheld EST is to be used, the appropriate software card should be used. You may use a generic handheld EST with heavy duty reader capability, but you will be limited to reading (scanning) the system.

EST-to-CA-to-chassis data connector

Operational electronically managed engine

PROCEDURE

The procedure will vary according to whether it is performed on an engine in chassis or one that is set up to run on a dynamometer or shop floor. Many colleges have engines set up to run out of chassis, and this may alter the means used to connect diagnostic instruments to the management electronics.

1. Access the appropriate software on the EST to be used for this exercise. If a generic handheld (HH) EST is to be used, insert the appropriate OEM software card or cartridge into the EST head. Locate the EST-to-chassis data bus connection hardware as shown in **Figure 15–4**. In a truck chassis, a data connector should be located on the left side of the steering column either around the lower dash or on the floor close to the driver's seat. If using a generic handheld EST, bear in mind that some systems will power up the unit directly through the data connector, while others will require separate power-up by means of a 12V auxiliary power source. Different cables are used for each type of system, so make sure that the right cable is used.

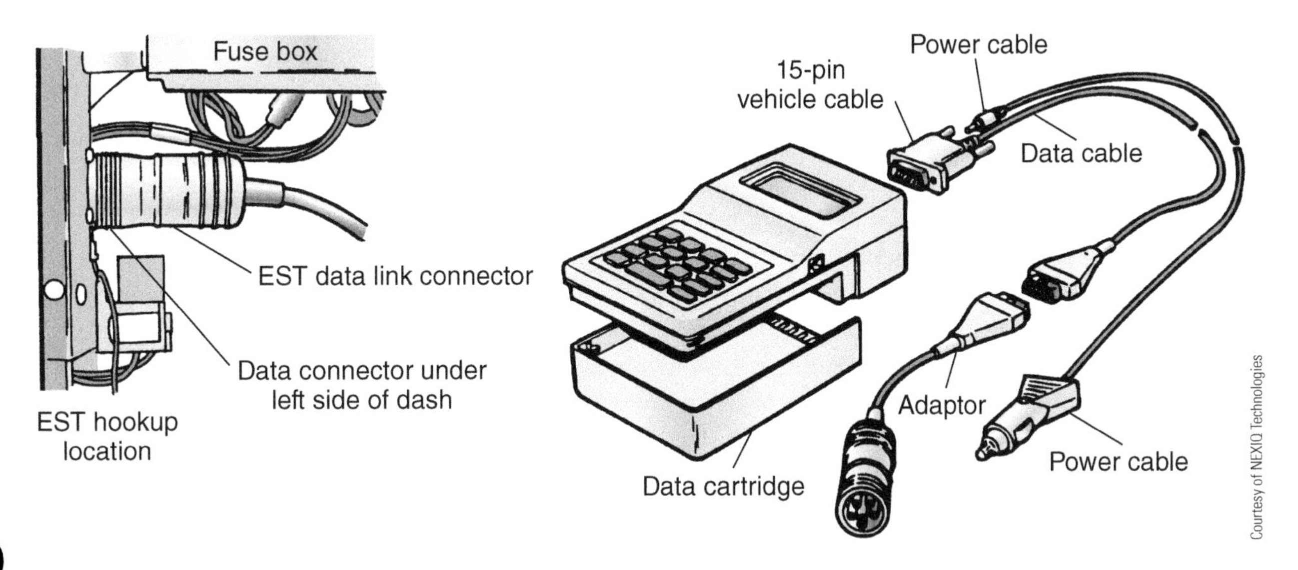

FIGURE 15–4 Handheld EST data connection.

Task completed ______________

2. Most generic handheld ESTs will power up as soon as they are connected to the vehicle electrical system. However, data is only displayed after the ignition circuit is switched on. Most PC-based ESTs require the ignition circuit to be switched on before data can be downloaded from the engine ECM.

   Task completed ______________

3. Obtain the entry download of data from the vehicle ECM to the EST. What you see in this step depends on whether there are multiple systems using the data bus. If there are multiple chassis electronic systems, you will have to select the engine controller SA #01 or MID #128. If the only on-board electronic system is the engine, you will automatically enter that SA/MID field. Most handheld ESTs have an LCD (liquid crystal display) window capable of reading only a limited number of lines of data. This means that the data displayed is a little more selective and you may have to scroll through data to find what you need to access. Whatever EST you are using, become familiar with it by scrolling through the various options and menus displayed on the screen. Identify system command keys, which are usually explained in the screen display. In a typical generic handheld EST, four directional arrow keys, a Function key, and an Enter key are used for navigating the data fields and system commands. No damage can result from scrolling data and option screens on ESTs, and this is an essential step in learning how to use the tool.

   Task completed ______________

4. Whatever EST you are using, after the system has been identified, a data list should appear. The presence of active and inactive codes logged in the ECM should be displayed. Scroll through the data list and make a note of anything unusual. In Nexiq HH ESTs, data lists are usually circular; that is, you can go forward or backward to any point.

   Task completed ______________

5. If there are no fault codes, try to create one by disconnecting an easy-to-access sensor such as a coolant level sensor. After a code has been logged, alter the code status from active to historic by reconnecting the sensor. Erase the historic code. In Nexiq ProLink iQ, try pressing the Function key. This readout will display engine and EST command functions; the default is engine functions. Engine functions display fault codes and enable the EST to perform tests and customer data programming. EST command functions relate to the way in which the EST itself functions, such as display contrast adjustment, serial port options (RS 232 port printer or PC connections), and snapshot mode.

   Task completed ______________

6. Program idle speed. This is a simple customer-data programming option. Certain requirements must be met before attempting to reprogram idle speed. Typically these are as follows:

   1. No active codes
   2. Vehicle stationary
   3. Clutch fully engaged
   4. Accelerator pedal at 0 travel position
   5. Parking brake applied

   Task completed ______________

7. Select the "Program idle speed" option. The screen will usually indicate the idle speed range window in rpm; typically this might be 500–750 rpm. This means that any idle speed value within this range can be programmed to the ECM. In a PC-based system, you will be prompted through the procedure step by step. In a generic handheld EST, the screen will suggest you press Enter to continue. The next screen will display "CURRENT rpm" and a value, followed by "DESIRED rpm" on the next line with the same value. Using the up and down arrow keys will change the displayed value on the "DESIRED rpm" line; it will not alter the rpm at this time. When the desired rpm is at the correct new value, press the Enter key. Depending on the system, this may effect an immediate change in rpm. Some systems may have to be shut down and restarted before the new idle rpm value is effected. This completes a simple exercise in customer data programming.

Task completed ______________

Instructor Check/Comments

____

____

____

JOB SHEET 15.6

Name ______________________ Date ______________

Job Description: Perform a snapshot test.

Performance Objective: After completing this assignment, the student should be able to troubleshoot an electronic management system by performing a snapshot test with the objective of analyzing data frames.

Text Reference: Chapter 15.

Protective Clothing: Required for shop exercise sessions: hearing, eye, hand, and foot protection, and shop coat/coveralls.

Tools and Materials:

PC station, palm unit, or generic handheld EST
OEM software card or download, or proprietary PC software
Electronically managed engine with J1708 or J1939 data link

PROCEDURE

Snapshot testing can be used as a troubleshooting strategy in most electronically managed systems. It can be especially useful in determining the cause of intermittent system faults that log codes that read out as INACTIVE at the time of analysis. Snapshot mode essentially captures data frames both before and after an event; the event may be a manual trigger or the specific moment that an ECM logs a fault code. The amount of data that can be retained when the EST is in snapshot mode varies with the system.

1. Connect the EST to the chassis data bus connector. For purposes of this description, a generic handheld EST will be used, but the procedure is similar when using a PC. Select "Command functions" and scroll to SNAPSHOT mode.

   Task completed ______________

2. The menu selection should now read QUICK TRIGGER. Scrolling will present other options, including TRIGGER SET-UP, DATA UPDATE RATE, and REVIEW SNAPSHOT.

   Task completed ______________

3. QUICK TRIGGER allows fast access to the snapshot mode. Selecting QUICK TRIGGER means that the EST will automatically choose the default operation.

   Task completed ______________

4. TRIGGER SET-UP permits the technician to specify what is to be used as the trigger to begin a snapshot data analysis sequence. The choices are:

   1. Any numeric key
   2. Any code
   3. Specific SPN or PID
   4. Specific SID

   Task completed ______________

5. When "Any numeric key" is selected, a generic handheld EST places a marker in the recording at the moment any of the numeric keys on the keypad is pressed. This is useful in attempting to identify the cause of a drivability or performance problem that does not set trouble codes.

   Task completed ______________

6. If "Any code" is selected, the EST will open a snapshot sequence the instant any trouble code is set.

Task completed ________________

7. If "Specific PID," or "Specific SPN," or "Specific SID" is selected, consult the OEM listing of PIDs, SPNs, and SIDs and select the field required.

Task completed ________________

8. Launch the EST into snapshot mode using either a manual trigger or any code trigger. Intentionally disable a sensor that will result in the logging of a fault code.

Task completed ________________

9. While the EST is waiting for a trigger to occur, the data display window will read WAITING FOR TRIGGER on the bottom line. As soon as the trigger is set, this will change to PROCESSING TRIGGER. The snapshot data recording process will terminate at any time a numeric keypad button is depressed. When the relevant data has been recorded, the data display window will change to enable the technician to interpret the data.

Task completed ________________

10. REVIEW DATA. This option plays back the snapshot data for analysis. The upper three lines display data. The bottom line contains the snapshot operating information. The letter *T* and the numeral that follows indicate the number of the frame that contains the trigger. To the right of the *T*, the letter *C* and the numeral that follows indicate the current frame being displayed. To the right of the lower line, GO TO . . . permits the technician to enter a number that will access a specific data frame without having to scroll through them all.

Task completed ________________

11. During analysis, to observe what codes exist, pressing Enter in the GO TO location will result in displaying the fault code(s).

Task completed ________________

12. Selecting "Data update rate" can alter the rate or intervals at which data is updated. The time that elapses between streams is known as the *delay*. The delay can be programmed at intervals of 0.1 to 9.9 seconds. Pressing Enter sets the delay period. Familiarity with snapshot mode troubleshooting can be a valuable asset in troubleshooting. Use both manual-trigger and fault code-triggered sequences to explore its potential.

Task completed ________________

Instructor Check/Comments

____

____

____

JOB SHEET 15.7

Name ______________________________ Date ______________

Job Description: Identify MIDs on a chassis data bus.

Performance Objective: After completing this assignment, the student should be able to identify the physical location of modules [source addresses (SAs) or message identifiers or MIDs] networked to a truck chassis data bus.

Text Reference: Chapter 15.

Protective Clothing: Required for shop exercise sessions: hearing, eye, hand, and foot protection, and shop coat/coveralls.

Tools and Materials:

Truck equipped with multiple networked ECMs
EST and appropriate connection to hardwire it to the chassis data bus
This worksheet and something to write with

PROCEDURE

1. Connect an EST to the chassis data bus. Identify and list all the SAs or MIDs by number and an alpha explanation (such as "engine: SA 01/MID 128").

Task completed ______________

2. Use **Figure 15–5** to indicate the physical location of each SA/MID you identified on the EST. Add any additional ones you may have located.

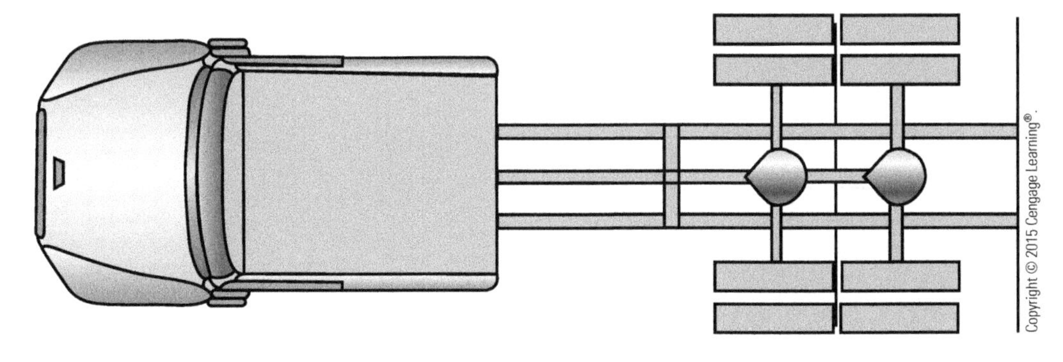

FIGURE 15–5 Identify the physical location of SAs and MIDs on a chassis data bus.

Task completed ______________

Instructor Check/Comments

___

___

___

INTERNET TASKS

Search the Internet for information on suppliers of vehicle computer systems, sensors, and actuators. In your research, you will discover that regardless of the OEM, the field is dominated by just a few specialty manufacturers who engineer systems onshore while having them manufactured offshore. Enter the following manufacturers into your search engine and try to determine what sort of processing power is used in the most recent vehicle engine controllers.

1. Motorola®
2. Siemens VDO®
3. Bosch®
4. Delphi®
5. Pye Electronics®

STUDY TIPS

Identify five key points in Chapter 15.

Key point 1 ______

Key point 2 ______

Key point 3 ______

Key point 4 ______

Key point 5 ______

CHAPTER 16

Electronic Diesel Fuel Injection Systems

OBJECTIVES

After studying this chapter, you should be able to:

- Describe the system layout and the primary components in current full-authority, electronic fuel management systems.
- Identify the key features of electronic unit injector (EUI) and common rail diesel fuel injection systems.
- Outline the role the four primary subsystems play in managing an EUI-fueled engine.
- Describe the operating principles of two-terminal (single-actuator) EUIs.
- Describe the operating principles of four-terminal (two-actuator) Delphi EUIs.
- Describe how the electronic/engine control module manages EUI duty cycle to control engine fueling.
- Outline some of the factors that govern the EUI fueling and engine output.
- Identify common rail diesel fuel systems.
- Identify some of the diesel engines currently using common rail diesel fuel injection.
- Trace fuel flow routing from tank to injector on common rail diesel-fueled engines.
- Describe the electronic management circuit components used in common rail fuel systems.
- Describe the operation of the inline and radial piston pumps used to achieve sufficient flow to produce rail and injection pressures in a typical CR system.
- Understand how rail pressures are managed in an electronically managed, common rail diesel fuel system.
- Outline the operation of an electrohydraulic injector.
- Identify some of the characteristics of different original equipment manufacturer common rail diesel fuel injection systems.
- Describe the operation of a amplified common rail system.

END OF CHAPTER REVIEW QUESTIONS

1. Which type of EUI was used by most OEMs before 2007?
   a. Twin actuator
   b. Two-terminal
   c. Four-terminal
   d. Mechanically switched
2. When EUIs are used on highway engines produced after 2007, which type of injector is required?
   a. Two-terminal EUI
   b. Two-terminal EUP
   c. Four-terminal EUI
   d. Four-actuator EUP
3. How many actuators are used in a four-terminal EUI?
   a. None
   b. One
   c. Two
   d. Four
4. Which of the following values would represent a typical NOP for a four-terminal EUI equipped with an EHI?
   a. Soft value NOPs controlled by ECM
   b. 3,800 psi
   c. 5,000 psi
   d. 22,000 psi
5. Technician A says that when a four-terminal EUI is injecting fuel into the engine cylinder, the SV actuator must be energized. Technician B says that for fuel injection to take place in a four-terminal EUI, the NCV actuator must be energized. Who is correct?
   a. Technician A
   b. Technician B
   c. Both A and B
   d. Neither A nor B
6. Which of the following best describes the type of injector used with CR fuel systems?
   a. Pintle injectors
   b. Electrohydraulic injectors
   c. Poppet injectors
   d. Electronic unit injectors
7. How many high-pressure pump elements are used on a Bosch radial piston pump fueling a six-cylinder engine with a CR fuel system?
   a. One
   b. Two
   c. Three
   d. Six
8. Which of the following are advantages of multipulse injection cycles?
   a. Lower cold-start emissions
   b. Lower noise levels
   c. Better fuel economy
   d. All of the above
9. Which of the following best describes the means by which the ECM drivers control the actuation of CR EHIs?
   a. V-Ref signal
   b. V-Bat signal
   c. Distributor spike
   d. Pulse width modulation

10. Which of the following is the key component used to signal actual rail pressure to the ECM?
    a. Rail pressure sensor
    b. Rail pressure control valve
    c. Pressure limiter valve
    d. Flow limiter valve
11. What device on a typical CR system prevents constant fueling of a cylinder if one of the EHIs sticks in the open position?
    a. Flow limiter
    b. Pressure limiter valve
    c. Pressure control valve
    d. Collapse of accumulator
12. What force is used to hold a CR EHI nozzle valve in its closed and seated position when the engine is running?
    a. Spring force only
    b. Electrical force only
    c. Combined hydraulic and electrical force
    d. Combined hydraulic and spring force

JOB SHEET 16.1

Name ______________________________ Date ______________

Job Description: Perform an electronic cylinder cutout test on an EUI-fueled engine.

Performance Objective: After completing this assignment, the student should be able to perform both automatic and manual-trigger EUI cylinder cutout tests. The tests will vary somewhat by OEM, but the general principle is the same.

Text Reference: Chapter 16.

Protective Clothing: Required for shop exercise sessions: hearing, eye, hand, and foot protection, and shop coat/coveralls.

Tools and Materials:

DDEC, CELECT Plus, Cat ADEM, or VECTRO-managed engine
PC and OEM software, or generic handheld EST (such as Nexiq iQ) and with the appropriate DDC software
EST to a communications adaptor (CA) and data bus (J1708 or J1939) connection hardware

PROCEDURE

The cylinder cutout test is a no-load method of assessing cylinder balance, performed using an EST with the appropriate OEM software or handheld EST. The bolded fields used in this worksheet correspond to the cues commonly used by the display interface. Most EUI management systems have similar EUI tests that replicate the one outlined here.

1. Connect the EST to the chassis data connector and select the Cylinder Cutout test mode. If the engine is in a vehicle, ensure that the transmission is in neutral and the parking brake system is fully applied.

   Task completed ______________

2. The screen menu will then ask whether a "new test" should be run or a previous test be "reviewed." The default (in brackets) is "new test," so when the Enter button is pushed, the next screen will present the options "automatic" or "manual" test. The default is automatic.

   Task completed ______________

3. The cylinder cutout test may be performed at idle rpm or at 1,000 rpm. The default is idle rpm, which has been proven to produce more accurate results.

   Task completed ______________

4. In automatic test mode, the EST controls the entire test, displaying the test data as it is logged. The engine is first run with all the EUIs operating. The resulting pulse width (PW) displayed is that required to make the engine run at the "command" rpm, either idle rpm or 1,000 rpm. This is known as **base PW**.

   Task completed ______________

5. After the base PW has been determined, each EUI is cut out in turn. This means that in order for the engine to continue to run at the command rpm, the PW will have to be increased, assuming the cutout EUI is operative. The PW that results from each cylinder cut out is displayed. The cylinder cutout test will continue until all the cylinders have been cut out. If a nonfunctioning EUI is cut out, then there will be NO increase in PW.

   Task completed ______________

6. In manual test mode, the cylinder cutout test is similar to the automatic test, but the technician selects the cylinders to be cut out. On a generic handheld EST, the Left and Right arrow keys are used to index the cylinder number. When Enter is pressed, the result of the cut is then displayed.

   Task completed ______________

7. Cylinder cutout test data can be analyzed in two ways. The test data can be reviewed by the EST in Review mode or the results can be printed out using a PC-driven printer or the RS 232 port and a printer on a handheld EST. Printed analyses enable a more complete overview of the engine conditions through a cutout test; this method is preferred.

   Task completed ______________

Instructor Check/Comments

____

____

____

JOB SHEET 16.2

Name ______________________________ Date ______________

Job Description: Perform wireless reprogramming of a set of EUI calibration codes on an EUI-fueled engine.

Performance Objective: After completing this assignment, the student should be able to make a wireless connection to an EUI-fueled engine and, using OEM software, reprogram a set of EUI calibration codes to the ECM.

Text Reference: Chapters 14, 15, and 16.

Protective Clothing: Required for shop exercise sessions: hearing, eye, hand, and foot protection, and shop coat/coveralls.

Tools and Materials:

EUI-fueled engine in an appropriate chassis, enabled for either hardwire or wireless connectivity.
Portable PC loaded with OEM software, and preferably, Internet access to an OEM data hub.

PROCEDURE

For this exercise, we will assume that you are working on a typical EUI-fueled engine, but remember that there are minor differences in the procedure used by each OEM. Begin by booting the notebook EST and launching the OEM software.

1. Establish and verify the hardwire or wireless connection with the truck chassis data bus.

   Task completed ______________

2. Launch the OEM software and ensure that it makes the required Internet connection.

   Task completed ______________

3. Enter the customer data programming field and select the injector calibration field.

   Task completed ______________

4. Identify the calibration code on each EUI. On Delphi E3 injectors (used in post-2007 engines), this is usually a four-digit number that appears on the side of the injector.

   Task completed ______________

5. Program each EUI calibration code into the appropriate field and save.

   Task completed ______________

6. Back out of the hardwire or wireless connection with the truck chassis data bus.

   Task completed ______________

Instructor Check/Comments

____

____

____

JOB SHEET 16.3

Name ______________________________ Date ____________________

Job Description: Familiarization with some general service procedures on CR-fueled diesel engines.

Performance Objective: After completing this assignment, the student should have a good understanding of the key components and some of the diagnostic routines on typical CR-fueled diesel engines.

Text Reference: Chapters 14, 15, and 16.

Protective Clothing: Required for shop exercise sessions: hearing, eye, hand, and foot protection, and shop coat/coveralls.

Tools and Materials:

Any one of the following CR engines:

Isuzu Duramax 6600 or 7800
Cummins ISB (after 1999), ISC (after 2004), ISL (after 2004), or ISX (after 2010)
Ford 6.7 liter engine (after 2011)
Any of the Paccar MX engine family after 2013
Caterpillar C7 or C9 (after 2007)

The OEM diagnostic software and online service information system (SIS).

PROCEDURE

At the time of this writing, there are four manufacturers of CR-fueled systems, and almost all engine OEMs of small-, medium-, and large-bore diesels are using CR fueling on at least some of their engines. This exercise is therefore general in nature. Students are cautioned not to remove any engine parts without first consulting the OEM service literature. Failure to observe this rule can result in costly repairs to keep the engine running.

1. Student to identify the engine family and fuel system manufacturer:
   - OEM ________________
   - Engine family ________________
   - Displacement ________________
   - Management system ________________ (Bosch/Cat/Denso, etc.)
   - Management electronics ________________ (IS, ADEM, DDEC, etc.)

Task completed ________________

2. Student to mark the location of the following components:
   - ECM ________________
   - High-pressure pump ________________
   - Note type of high-pressure pump ________________
   - Rail pressure management control (RPMC) valve ________________
   - Rail pressure sensor ________________
   - Crankshaft speed sensor ________________
   - Camshaft speed sensor ________________
   - Accelerator pedal sensor ________________

- Accelerator pedal sensor signal type (Pot/Hall) ________________
- EGR actuator ________________
- EGR MAF sensor ________________
- Boost pressure sensor ________________
- Coolant temperature sensor ________________
- Oil pressure sensor ________________

Task completed ________________

3. Student to measure and record system voltage using a DMM or access data bus and scan reading ________________ V-DC.

Task completed ________________

4. Student to measure and record V-Ref system voltage using a DMM (separate sensor connector) ________________ V-DC.

Task completed ________________

5. Student to identify data bus used to network engine to chassis electronics:
   (circle appropriate choice) J1850/CAN-C J1587/1708 J1939

Task completed ________________

6. Student to identify the EHIs used with system. **Caution:** Do not attempt to pull the injectors. Use the OEM SIS to obtain the information if you can.
   - EHI manufacturer ________________
   - EHI actuator type ________________ (solenoid/piezo)
   - EHI voltage peak ________________
   - EHI current draw ________________
   - Electrically disconnect each EHI and check actuator resistance:

   Inj. 1___ Ω, Inj. 2___ Ω, Inj. 3___ Ω, Inj. 4___ Ω, Inj. 5___ Ω, Inj. 6___ Ω

Instructor Check/Comments

____

____

____

INTERNET TASKS

Pick a CR-fueled engine you want to know more about, and search the Internet for information on it. You will find it easy to get as much data as you wish on CR-fueled engines because CR is becoming the dominant diesel fuel-management system. Enter the following key words into your search engine and see what you come up with.

1. Bosch® CR
2. Caterpillar® CR
3. Denso® CR
4. Navistar® Maxxforce™
5. Paccar® DAF CR systems

STUDY TIPS

Identify five key points in Chapter 16.

Key point 1 ______________________________

Key point 2 ______________________________

Key point 3 ______________________________

Key point 4 ______________________________

Key point 5 ______________________________

CHAPTER 17

Emissions

OBJECTIVES

After studying this chapter, you should be able to:

- Define the origin of the word *smog*.
- Define photochemical smog and describe the conditions required to create it.
- Identify some common tailpipe emissions.
- Outline the operating principles of C-EGR, oxidation catalytic converters, reduction catalytic converters, diesel particulate filters, and selective catalytic reduction (SCR).
- Explain how an opacity meter functions.
- Describe the SAE J1667 test procedure.
- Analyze diesel engine smoke.

END OF CHAPTER REVIEW QUESTIONS

1. Which of the following compounds has recently been classified by the U.S. Supreme Court as a pollutant?
   a. Oxides of nitrogen
   b. Hydrocarbons
   c. Sulfur dioxide
   d. Carbon dioxide
2. Which of the following is present in the largest quantity by weight in the diesel engine cylinder during combustion?
   a. Oxygen
   b. Nitrogen
   c. Fuel
   d. Carbon monoxide
3. Which tailpipe emission does an NO_x adsorber catalyst attempt to reduce?
   a. Particulate matter (PM)
   b. Oxides of nitrogen
   c. Hydrocarbons
   d. Ozone
4. Which of the following exhaust gas test instruments uses a light extinction principle to measure smoke density?
   a. CVA sampler
   b. Seven-gas analyzer
   c. Two-gas analyzer
   d. Opacity meter

5. In which of the following forms is oxygen most commonly found at ground level?
   a. O
   b. O_2
   c. O_3
   d. O_4
6. Which of the following results from "perfect" combustion of an HC fuel when it is burned with oxygen?
   a. Carbon monoxide and carbon dioxide
   b. Carbon dioxide and nitrogen dioxide
   c. Nitrogen dioxide and water
   d. Carbon dioxide and water
7. When a diesel engine is operated at lower-than-normal temperatures, which of the following tailpipe emissions is likely to increase?
   a. Oxides of nitrogen
   b. Carbon dioxide
   c. Ozone
   d. Hydrocarbons
8. When a diesel engine is operated at higher-than-normal temperatures, which of the following tailpipe emissions is likely to increase?
   a. Oxides of nitrogen
   b. Carbon dioxide
   c. Ozone
   d. Hydrocarbons
9. Which of the following compounds is classified as a greenhouse gas responsible for contributing to global warming?
   a. Oxides of nitrogen
   b. Hydrocarbons
   c. Carbon dioxide
   d. Sulfur dioxide
10. Which of the following describes what happens in an oxidation-type catalytic converter?
    a. Nitrogen in NO_x is burned.
    b. CO and HCs are burned.
    c. HCs are filtered out.
    d. PM is filtered out.

JOB SHEET 17.1

Name ______________________________ Date ______________

Job Description: Perform a J1667 modeled smoke test using an opacity meter.

Performance Objective: After completing this assignment, the student should be able to understand the procedure behind J1667 testing and perform the test.

Text Reference: Chapter 17.

Protective Clothing: Required for shop exercise sessions: hearing, eye, hand, foot protection, and shop coat/coveralls.

Tools and Materials:

Functional highway diesel engine
PC-driven opacity meter calibrated for state or provincial J1667 testing
OEM diagnostic software and online service information system (SIS)

PROCEDURE

The procedure will vary depending on the equipment you are using and whether the station is set up to officially report results to the state or provincial database. The exercise can be performed just as effectively using non-official light extinction testing apparatus. However, equipment that is set up for official testing will enforce the test preconditions and prompt each throttle snap. **Figure 17–1** shows a typical setup for a J1667 test.

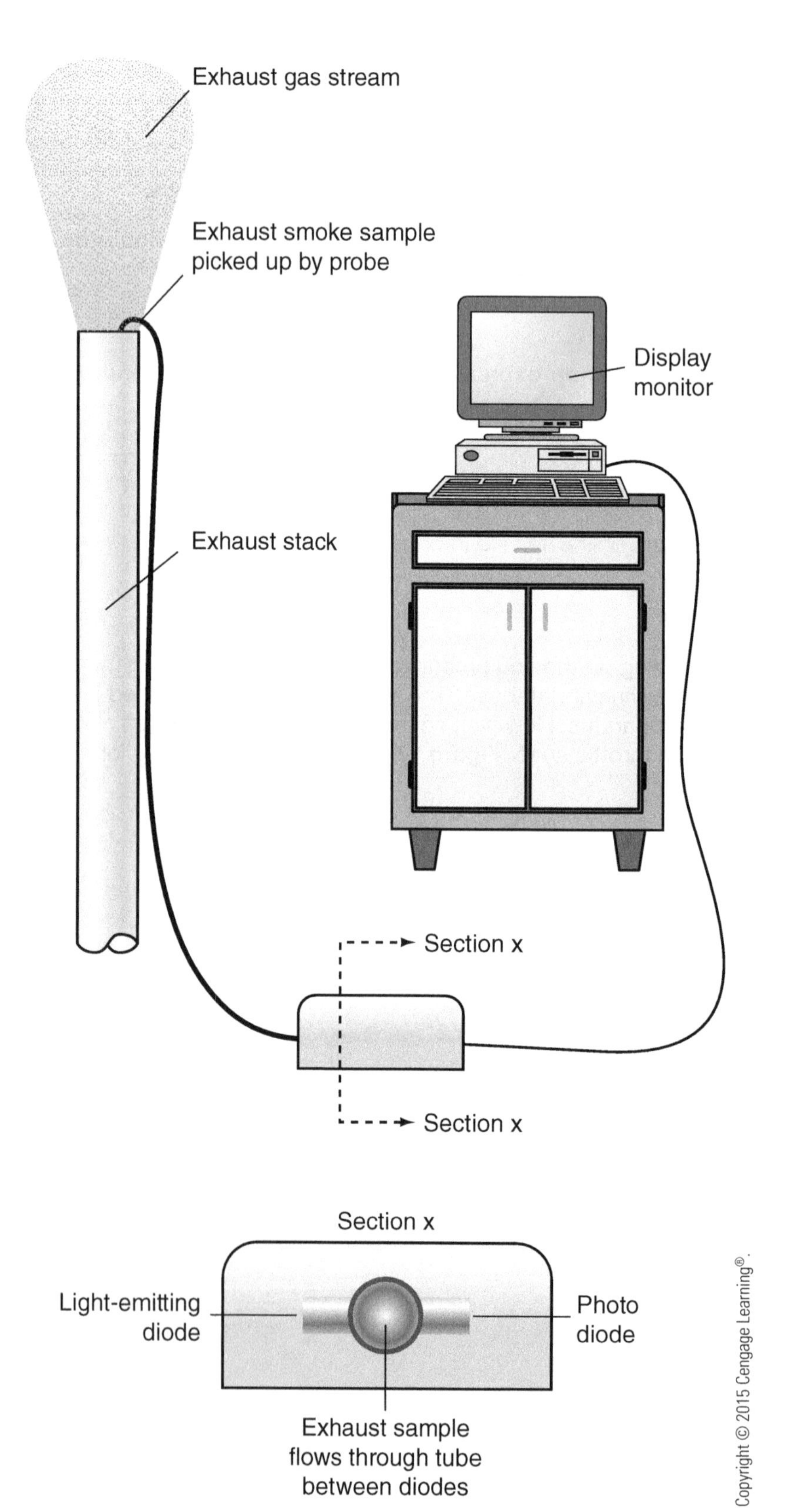

FIGURE 17–1 Setup for a J1667 opacity test that uses light extinction technology.

1. Meet the test preconditions:
   - Engine oil temperature must exceed 150°F ____________.
   - Parking brakes released; hold-off psi +90 psi ____________.
   - Temperature above dew point and between 35°F and 86°F ____________.
   - Open file for the chassis to be tested in opacity meter software ____________.
   - Enter engine data and exhaust stack dimension data ____________.

   Tasks completed ____________

2. Preliminary snaps. Three full snaps are required to clear the exhaust system of loose particles and precondition the vehicle. Engine rpm must drop to the specified low idle speed after each snap.

   Task completed ____________

3. Official snaps. Three official snaps should be undertaken as prompted by test instrument software. Accelerator is snapped to high idle and held for one to four seconds or until prompted by the opacity meter to release. Engine rpm must be allowed to drop to the specified low idle speed; it must then run at the specified low idle for a minimum of five seconds and a maximum of 45 seconds before initiating the next snap as prompted by the opacity meter software.

   Task completed ____________

4. Validation of the official test results. The difference between the highest and the lowest maximum opacity readings of the three official test snaps should be within five opacity percentage points. If the difference is less than 5 percent, the meter software computes the average. If the difference is greater than 5 percent, additional official snaps must be undertaken, up to a maximum of nine.

   Task completed ____________

5. Interpret drift factor. Validation is required that the drift between three official test cycles does not exceed 2 percent opacity. A variation greater than 2 percent will result in an invalid test.

   Task completed ____________

Instructor Check/Comments

____

____

____

INTERNET TASKS

Check out the different models of J1667 testing used by different jurisdictions. Because smoking diesel engines are somewhat of a political issue, you will find the actual standards required to pass vary somewhat. Enter the following key words into your search engine and see what you come up with.

1. California CARB diesel
2. Drive clean testing diesels
3. dieselnet.net

STUDY TIPS

Identify five key points in Chapter 17.

Key point 1 ______________________________

Key point 2 ______________________________

Key point 3 ______________________________

Key point 4 ______________________________

Key point 5 ______________________________